Radio Frequency Identification Engineering

Radio frequency identification (RFID) has become an undeniable aspect of modern living, being used from logistics, access control, and electronic payment systems to artificial intelligence, and as a key building block of the internet of things. Presenting a unique coverage of RFID reader design and engineering, this is a valuable resource for engineers and researchers, aiding in their mission of fulfilling current and future demands in the RFID space.

Providing a cohesive compilation of technical resources for full-stack engineering of RFID readers, the book includes step-by-step techniques, algorithms, and source code that can be incorporated in custom designs. Readers are invited to explore the design of RFID interrogators based on software-defined radio for flexible, upgradeable solutions as well as low-complexity techniques for engineering low-cost RFID readers. Additionally, the authors provide insight into related topics such as waveform design optimization for improved reading range and novel quadrature backscatter modulation techniques.

Alírio Soares Boaventura is a senior research scientist at Maybell Quantum Industries in Denver, Colorado, and worked previously as a research scientist for over five years at the National Institute of Standards and Technology in Boulder, Colorado. He is the recipient of an IEEE MTT-S graduate fellowship award, co-recipient of the 2011 URSI-Portugal/ANACOM Prize, and was featured in the publication *University of Aveiro, 40 Years, 40 Inventors, 40 Entrepreneurs*.

Nuno Borges Carvalho is currently a full professor and a senior research scientist with the Institute of Telecommunications, University of Aveiro, and an IEEE Fellow. He coauthored *Intermodulation Distortion in Microwave and Wireless Circuits* (Artech House, 2003), *Microwave and Wireless Measurement Techniques* (Cambridge University Press, 2013), *White Space Communication Technologies* (Cambridge University Press, 2014), and *Wireless Power Transmission for Sustainable Electronics* (2020). He is a distinguished lecturer for the RFID Council and was the 2023 President of the IEEE-MTT Society.

"This book is a must-have reference handbook for researchers aspiring to design and build a versatile functional software-defined RFID reader. It is also one of the best resources for anyone looking to learn more about the world of RAIN RFID. This highly recommended book covers a wide range of topics, starting from spectrum, standards, backscatter technology definitions, and all the way to reader architecture choices and RFID signal processing, including encoding, modulation, and filtering. With sections on advanced research concepts such as multicarrier systems or adaptive self-jamming suppression, as well as many exercises and practical MATLAB code examples, it can be used as a great textbook for students."

Pavel Nikitin, University of Washington

"This book provides a comprehensive and thorough coverage of existing Radio Frequency Identification (RFID) technology and current and future areas of research and development. The book presents the wide variety of concepts foundational to RFID in an accessible manner, providing in-depth information about each concept."

Peter Hawrylak, The University of Tulsa

Radio Frequency Identification Engineering

How to Engineer an RFID Reader

ALÍRIO SOARES BOAVENTURA
University of Aveiro, Portugal

NUNO BORGES CARVALHO
University of Aveiro, Portugal

CAMBRIDGE
UNIVERSITY PRESS

Shaftesbury Road, Cambridge CB2 8EA, United Kingdom

One Liberty Plaza, 20th Floor, New York, NY 10006, USA

477 Williamstown Road, Port Melbourne, VIC 3207, Australia

314–321, 3rd Floor, Plot 3, Splendor Forum, Jasola District Centre, New Delhi – 110025, India

103 Penang Road, #05-06/07, Visioncrest Commercial, Singapore 238467

Cambridge University Press is part of Cambridge University Press & Assessment, a department of the University of Cambridge.

We share the University's mission to contribute to society through the pursuit of education, learning and research at the highest international levels of excellence.

www.cambridge.org
Information on this title: www.cambridge.org/9781108489713

DOI: 10.1017/9781108779265

When citing this work, please include a reference to the DOI 10.1017/9781108779265

First published 2025

A catalogue record for this publication is available from the British Library.

A Cataloging-in-Publication data record for this book is available from the Library of Congress.

ISBN 978-1-108-48971-3 Hardback

Additional resources for this publication at www.cambridge.org/9781108489713

Alírio: "Dedico este livro aos meus pais Alfredo e Maria Boaventura e aos meus irmãos Luiza, Rosa e José."

Nuno Borges: "Dedico este livro à minha esposa Raquel Madureira e aos meus filhos Rebeca, Tomás e Sara."

Contents

Foreword

Passive wireless systems are in high demand today because they allow for the implementation of inexpensive, low-complexity, and high-performance solutions for ubiquitous remote sensing and communication perfectly adapted to the environment. Such systems are made possible by exploiting two capabilities of electromagnetic waves, namely wireless communication as first demonstrated by Guglielmo Marconi, and wireless power transfer as first proposed by Nikola Tesla. The exploitation of these capabilities during the first decades of the twentieth century gave rise to many applications, the most popular ones being the broadcasting of radio and television shows. Also, World War II worked as a catalyst for several military applications, the most emblematic of which being radar and IFF (identify friend or foe), which are based on communication by means of reflected waves.

During the same period, some pioneering works were dedicated to coupling information and power in the same signal, opening the door to passive wireless systems. An early application of this concept consisted of an ingenious device developed by Leon Theremin, which was used as a spying microphone, and became known as "the thing." "The thing" is certainly the ancestor of modern RFID tags. The effective combination of wireless power and information, and adoption of the passive wireless concept, were enabled by the development of electronics and the birth of microelectronics following the invention of transistors around 1947. Among practical applications of this concept, Radio Frequency Identification (RFID) is probably the most relevant. Today, there is a growing effort in the research community to expand the passive RFID concept beyond identification to include environmental sensing, data logging, and advanced processing capabilities. By turning RFID tags into RFID sensors, augmented RFID provides more than identification and enables advanced concepts such as the internet of things (IoT) and artificial intelligence assisted RFID.

In recent decades, RFID users have grown significantly from a relatively small circle of engineers and electronics researchers to industry, commerce, and services. This evolution has been made possible thanks to the development of regulation and standardization rules, which allow for the interoperability of devices and subsystems from different manufacturers. Such standards concern both signal characteristics and communication protocols as well as their informational content. The progress made in UHF CMOS RFID chip technology, in particular the large scale of digital and RF co-integration, has allowed a huge reduction in the cost of RFID tags, making them cost effective for item-level tagging. Indeed, the number of deployed tags went from

4 billion in 2015 to 20 billion in 2020 and continues to grow at a compound annual growth rate of 10%.

To meet current demand and allow for effective implementations, the use of standardized and reconfigurable radio solutions is to be preferred. The software-defined radio paradigm, introduced by Joseph Mitola in the 1990s and applied in this book, is well suited to this purpose.

While there is an abundant literature on UHF RFID tags and their operation, the same cannot be said for the UHF RFID reader or interrogator which is a key element of ubiquitous IoT. The authors of this book have set themselves a two-fold objective: (i) meet literature needs for UHF RFID reader design by providing a compilation of technical resources, and (ii) provide practical implementations of UHF RFID readers using software-defined radio platforms. The authors detail diagrams, propose algorithms, and provide source code to implement ISO 18000-63 functions. The book presents a rich collection of experiments and practical implementations. It also covers other IoT-relevant topics such as wireless low-power wake-up radios and novel backscatter modulation schemes.

This book is authored by two well-known scientists with several years of experience and expertise in RFID and SDR technologies. It is an excellent resource for the application of RFID and SDR solutions.

Smail Tedjini

Emeritus Professor, Grenoble-INP, Université Grenoble Alpes (UGA)
Life Fellow, Institute of Electrical and Electronics Engineers (IEEE)
Fellow, International Union of Radio Science (URSI)

Acknowledgments

To all those who directly or indirectly contributed to this work, including our families, friends, and colleagues who supported us and the authors whose works have inspired us.

Special thanks to Smail Tedjini, Ricardo Gonçalves, Edward Keehr, André Prata, Joao De Deus Luz, Ricardo Fernandes, Renato Graça, Apostolos Georgiadis, Anna Collado, and Gustavo Avolio.

Thanks to the Department of Electronics and Telecommunications at the University of Aveiro and its faculty, and the Portuguese Science and Technology Foundation and the Institute of Telecommunications for the institutional and financial support.

1 Introduction

1.1 Background and Scope of the Book

The development of 5G and the internet of things (IoT) has enabled a connected environment with tens of billions of devices worldwide. This number is likely to amount to one trillion or more in the not-too-distant future where conventional devices like smartphones and laptops are expected to be outnumbered by non-electronic and non-cellular connected objects. Effective deployment and connectivity to these objects in ubiquitous IoT networks will require inexpensive, low-complexity, and low-maintenance radio technologies.

Radio frequency identification (RFID) and related technologies like backscatter and wireless power transfer are strong candidates for meeting these requirements. Even though the RFID concept has been around for several decades, it has only recently gained increased attention with the development of IoT, and while there is an abundant literature on the fundamentals and applications of RFID, there are fewer technical resources focused on a key element of RFID systems – the RFID reader.

This book helps to close this gap by providing a compilation of resources for full-stack RFID reader design and engineering spanning basic concepts, simulations, measurements, hardware, and software, and contains a rich collection of experiments and exercises. The book can be helpful for students, practitioners, and researchers looking to gain insight into the operating principles and engineering of UHF RFID readers and other similar systems. For entry-level readers, we provide the foundations for engineering functional systems. For experienced practitioners and researchers, we offer many resources to help overcome engineering problems and jump-start more advanced designs and research. The reader of this book should be familiar with signal processing concepts, digital modulation techniques, communication protocols, and radio system architectures.

There are 11 chapters, organized as follows: Chapter 1 walks the reader through the evolution of RFID technologies from the early days of radio transmissions in the nineteenth century to the internet of things. Chapter 2 introduces the fundamental concepts, applications, standards, and operating principles of RFID and offers a glimpse into design considerations and architectures of modern UHF RFID readers. Chapter 3 discusses the fundamentals of backscatter radio communications and presents recent developments, including novel quadrature backscatter modulation techniques for

IoT-RFID. Chapter 4 gives an overview of the ISO 18000-63 communication protocol, including data encoding, modulation, and storage, and presents real examples of reader-to-transponder transactions. Chapter 5 discusses the implementation of ISO 18000-63 downlink and uplink communication chains and provides practical algorithms and code for evaluating RFID signal processing chains. Chapter 6 explores low-complexity techniques for the design of a low-cost ISO 18000-63 RFID reader and presents a full-stack implementation to validate the proposed concepts. Chapter 7 presents an RFID reader design based on software-defined radio (SDR), including hardware and software implementations, and a demonstration of ISO 18000-63 operation in continuous-wave mode and a novel multicarrier mode. Chapter 8 provides a comprehensive discussion of self-jamming in passive-backscatter systems through the study of various self-jamming mitigation and suppression approaches, including some used in commercial off-the-shelf and integrated readers. Chapter 9 addresses wake-up radios for IoT applications and describes a custom implementation for wireless sensor networks. Chapter 10 evaluates the application of multicarrier signals to improve the efficiency of wireless power transfer systems and proposes efficient wireless power transmitter architectures, including a mode-locked active antenna array. Chapter 11 concludes the book with an elegant demonstration of wireless power transfer and backscatter applied to home automation through the design and prototyping of a battery-free remote control system and its integration into a TV receiver. In summary, this book covers the following topics:

- Basic concepts and applications of RFID
- Fundamentals and recent advances in backscatter communications
- Novel quadrature backscatter modulation techniques
- Practical digital signal processing for RFID applications
- Simple, low-cost UHF RFID reader implementation
- Software-defined radio UHF RFID reader design
- Self-jamming suppression in backscatter systems
- Low-power wake-up radios
- Unconventional wireless power transmission
- An RFID-based battery-less remote control system.

The concepts discussed in this book have received multiple recognitions, including an IEEE MTT-S microwave engineering graduate fellowship award, two nominations for the IEEE IMS best student paper, a best paper at the WPT conference, and an URSI Portuguese section and Portuguese ANACOM prize.

1.2   History of RFID Technology – From Early Radios to Today's IoT

This section gives a historical perspective of the evolution of RFID technology, from its inception with early radio experiments for information and power transmission to remote object detection and identification and its present-day maturity and role in the IoT.

1.2.1   Early Radio Experiments

Radio frequency identification shares a common history with radio-frequency communications, a journey that for many began in the nineteenth century with an important discovery made by the self-educated British scientist Michael Faraday. In 1831, Faraday reasoned that the movement of a magnet next to a conductor induces an electric current in the circuit, and established the law of electromagnetic induction [1]. Faraday also envisaged the existence of "lines of force" around electric charges and magnetic poles, and predicted the existence of electromagnetic (EM) waves [2]. These discoveries laid the foundations for the modern theory of electromagnetism.

A few decades later (in 1864), James Clerk Maxwell, a Scottish mathematician and scientist, validated Faraday's hypothesis on the existence of electromagnetic waves. In a volume entitled *A Dynamical Theory of the Electromagnetic Field*, Maxwell not only summarized the concepts of electricity known at that time, but also provided a unified theory describing electricity, magnetism, and light as manifestations of the same phenomenon [3]. He postulated that EM waves had the same characteristics as light, including propagation speed, and that light itself was a manifestation of the electromagnetic phenomenon.

In response to a contest launched by the Academy of Science in Berlin to experimentally validate aspects of the Faraday–Maxwell theory, Heinrich Hertz provided the ultimate proof of the existence of EM waves and their similarities with light, and became the first scientist to ever generate, transmit, and detect EM waves [2]. Hertz's experimental setup consisted of a tuned spark-gap transmitter and a tuned spark-gap receiver [4]. The transmitting oscillator comprised an induction coil, a Leyden jar used as a "condenser," and several accumulators. Oscillations were produced by discharging an electric condenser through an air gap, which produced a spark over the gap and originated an oscillatory phenomenon. The oscillatory wave was then radiated using metal plates and detected some meters away from the transmitter by means of a second coil with an air gap, where sparks were induced when the transmitter circuit was operated. The tiny sparks at the receiver were visually detected in a dark room. Hertz also used parabolic-shaped metal reflectors as mirrors to focus the electromagnetic beam. Despite his great findings, Hertz did not get interested in any practical application of EM waves. In the 1900s, Guglielmo Marconi, a visionary engineer and entrepreneur, introduced wireless telegraphy and made it a big commercial success.

1.2.2   Power Transfer via Radio Waves

Wireless power transfer is the key enabling technology of passive RFID, where transponders have no onboard battery and are powered remotely via radio waves. The first experiments with EM waves for wireless energy transfer are attributed to Nikola Tesla and date back to the late nineteenth and early twentieth centuries [5–8]. In 1893, Tesla demonstrated wireless lamps at the world's Columbian exposition in Chicago [9]. Tesla also worked on high-frequency and high-voltage transformers, and began to build the Wardenclyffe tower, a radiation station intended for wireless transmission of electricity [2]. Even though his early experiments on wireless electricity were hindered by technological limitations [10] and wireless power transfer activities were suspended for decades, the pioneer work of Tesla [11] played a key role in the twentieth-century technology revolution.

Leveraging technological advances that occurred during World War II, principally on high-power devices (e.g., magnetron microwave tubes), William C. Brown and others resumed wireless power transfer experiments in the 1960s. Brown proposed the concept of microwave power transmission [7, 12], introduced the rectenna, a rectifying antenna for receiving and rectifying microwaves [12], and carried out many microwave experiments in the 2.45 GHz band [8]. In 1964, he demonstrated the microwave power transmission concept by powering a subscale helicopter from the ground using a microwave beam at 2.45 GHz [13]. Following Brown's work, several other concepts and experiments were reported, including solar power satellites [14–16], microwave-powered airplanes [17], wireless vehicle charging, and inductive resonant coupling [18, 19].

As new consortiums, startups, and companies dedicated to wireless power transfer have emerged in recent years, this technology has increasingly become part of many consumer electronic products [20–24]. Some of the most successful applications of the wireless power transfer concept today include passive RFID and inductive charging of consumer electronic devices. Figure 1.1 shows a timeline of the concurrent history of radio communications, wireless power transfer, and RFID.

1.2.3   Remote Object Identification Using Radio Waves

One of the first uses of radio waves for remote sensing is attributed to John Baird, who received one of the earliest patents on radio object detection in 1926 [25]. Baird's concept consisted of remotely illuminating an object with a directional beam of radio waves and capturing the reflected signal to form an image of the target object. In 1935, Robert Watson-Watt developed the first practical radio detection and ranging (radar) system [26], which played a decisive role in World War II. By transmitting radio pulses and measuring their echoes, the system was able to remotely detect the presence of distant aircraft and ships.

A major limitation in early radar systems relates to their inability to distinguish between allied and adversary aircraft. This was overcome with the introduction of the identify friend or foe (IFF) system, where allied aircraft were equipped with transponder or responder devices that replied to the ground interrogation station with "friendly signatures." This can be regarded as the first system of true identification via radio frequency.

During the Cold War, Russian inventor Leon Theremin developed a device based on radio-frequency backscatter that allowed eavesdropping on conversations at remote locations [27–29]. This device comprised a monopole antenna attached to a resonant cavity with a sound-sensitive membrane (see Figure 1.2). Deformations on the membrane induced by acoustic waves caused the resonant frequency of the cavity and impedance presented to the antenna to change. When illuminated with a remotely

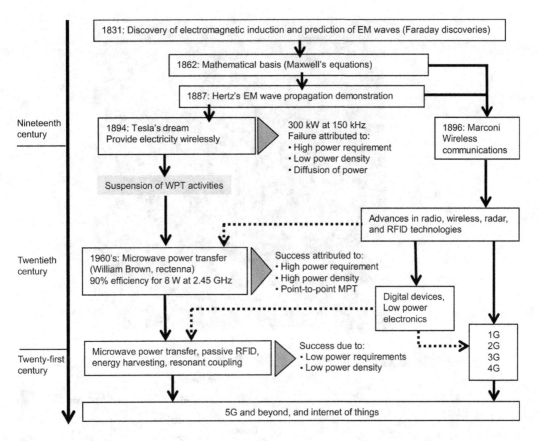

Figure 1.1 Concurrent history of radio communications, wireless power transfer, and RFID. Adapted from [10].

generated radio signal, the antenna retransmitted part of that signal modulated by the sound waves picked up by the membrane. This ingenious system was well ahead of its time when compared to early one-bit RFID electronic surveillance systems (EAS) developed in the 1960s, or even some *N*-bit transponder systems developed in the next decade.

In a paper entitled "Communications by Means of Reflected Power" published in 1948 [30], Henry Stockman theorized on the use of acoustic and electromagnetic wave reflection for passive communications, which forms the basis for modern back-scatter RFID systems. Since the early days of radio technology, researchers have explored approaches to enable battery-free operation of radio transceivers. In 1960, Donald B. Harris proposed a battery-free backscatter system for voice communication [31] in one the first attempts to combine wireless power transfer and electromagnetic wave backscatter, leading to the conception of modern passive RFID.

In the 1960s there was significant progress in RFID; the technology was for the first time used in commercial applications with counter-theft one-bit electronic surveillance systems becoming the first large-scale commercial success [32]. In 1966, Sensormatic

(a)

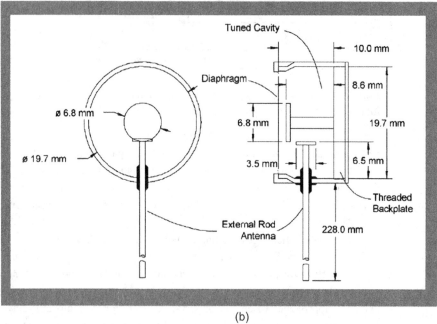

(b)

Figure 1.2 (a) Copy of the great seal of the United States with a hidden passive-listening device. (b) Diagram of the passive-backscatter listening device. Reprinted from [29].

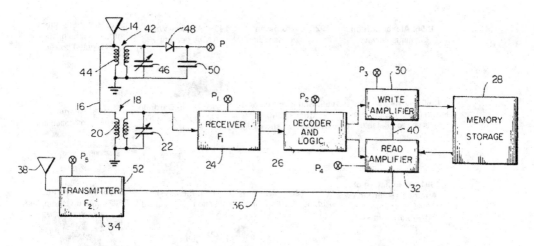

Figure 1.3 Drawing of the first patented passive RFID transponder with readable and rewritable memory. Reprinted from [33].

and Checkpoint emerged as the first RFID companies. Reportedly, the first RFID patent [33] was granted to Mario W. Cardullo in 1973 for a passive RFID transponder with rewritable memory and processing capability (Figure 1.3), originally meant for electronic toll collection. For the first time, the RFID transponder was proposed as a data carrier that could be remotely accessed by an interrogator. Following Cardullo's patent, many others were granted on similar concepts.

In the 1990s, the RFID transponder concept gained traction, primarily because of the large-scale deployment of electronic toll collection systems in several countries. During this decade, there was an effort on standardization to provide interoperability between systems and vendors, which was lacking in the earlier implementations. One of the first RFID standards to emerge was automatic equipment identification (AEI), approved by the Association of American Railroads (AAR) in 1991 to replace a previous optical barcode identification system [34]. In the following years, the AAR mandated the adoption of RFID transponders under the AEI standard to identify and track railcars.

In 1999, the MIT Auto-ID Center (funded by the Uniform Code Council and a consortium of global consumer product manufacturers) was founded and tasked with development of the electronic product code (EPC), a global RFID-based identification system intended to replace barcodes. The first EPC version was released in 2004, and one year later the International Standards Organization (ISO) incorporated the EPCglobal Gen 2 standard [35] into the ISO/IEC 18000-6 standard. Today, ISO 18000-63 is the most prevalent standard for passive radio frequency identification in the UHF band. Key aspects of this standard are summarized in Chapter 4 and further discussed in Chapter 5. Other key milestones in the development of RFID technology include the historic mandate by Walmart to all its suppliers to use RFID in their pallets and cases. The major milestones in the history of RFID technology are summarized in Figure 1.4.

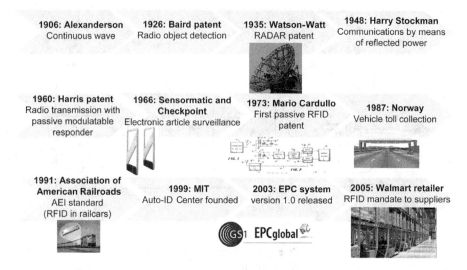

Figure 1.4 A timeline of RFID technology evolution.

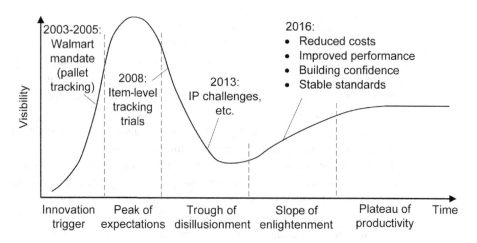

Figure 1.5 Simplified Gartner's RFID hype cycle.

1.2.4   A Turning Point

As with many emerging technologies, following an initial period of excitement and hyped expectations, progress in RFID stagnated, and the technology eventually went through a "trough of disillusionment" (Figure 1.5). In 2003, Walmart, then the world's largest retailer, mandated that all suppliers should tag their pallets and cases by 2005, which acted as a major "innovation trigger" and accelerated the introduction of RFID to replace barcodes; between 2005 and 2007 the technology reached its "peak of inflated expectations."

Given the limited success of item-level tracking trials in 2008, and after Walmart dropped its mandate a year later, developments in RFID technology stagnated.

Between 2013 and 2015, intellectual property (IP) challenges and overstretched roll-outs led RFID technology to its trough of disillusionment. But subsequent reductions in production costs, increasingly stable standards, and increasing maturity and performance eventually led the technology out of this stagnation phase. Today, RFID is an undeniable part of our personal and business lives and stands as a strong enabling technology candidate for the internet of things.

1.3   The Role of RFID in the Connected Future

The internet of things holds the promise of connecting people, machines, and every-day objects anywhere and anytime at an unprecedented scale. More than 50 billion connected devices are expected to be deployed worldwide by 2020 [36, 37], and beyond a trillion will be deployed in the not-too-distant future [38]. While most of today's connected devices are smartphones, tablets, laptops, and the like, they will soon be outnumbered by non-cellular and non-electronic everyday connected objects, which are expected to account for more than half of all internet-connected things by 2020 [39].

While it is still not entirely clear what the future internet of things will look like, the diverse range of potential applications and requirements makes it difficult, if not impossible, to conceive a one-size-fits-all radio technology to support the internet of things. But based on data throughput, cost, complexity, and power consumption criteria, one can currently identify two distinct internet segments: the internet of powered devices and the internet of everyday things (see Figure 1.6).

1.3.1   The Internet of Battery-Powered Devices

Besides personal mobile and portable wireless communications, potential applications in this segment of the internet include smart transportation, smart homes and smart cities, wearables, fleet management, smart luggage, security and surveillance, smart grids, facility management, and so on. These applications typically use battery-powered wireless devices of high complexity, high power consumption, and large bandwidths, capable of supporting data-intensive services (e.g., video streaming). IoT-powered devices are expected to amount to tens or hundreds of millions and will rely on short-range technologies like Wi-Fi (IEEE 802.11 wireless local area network (WLAN)), ZigBee (IEEE 802.15.4), and Bluetooth or wide-area cellular technologies such as 5G/6G.

These technologies typically exhibit high power consumption and there has been an effort to devise lower-power alternatives that can meet the requirements of IoT. Examples include Bluetooth low-energy, wakeup-based Wi-Fi (IEEE 802.11ba [39]), Sigfox, LoRa, and Ingenu. While wireless-powered devices will form the core of the future internet of things, they most likely will not be suitable for applications where cost, complexity, footprint, and consumption are critical.

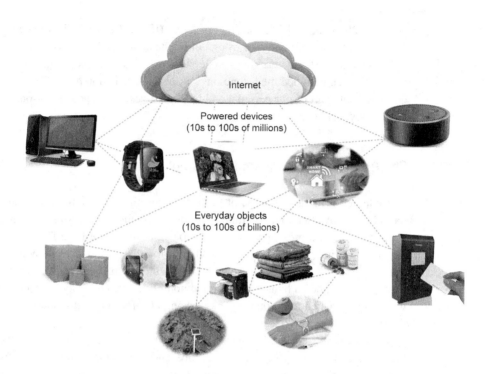

Figure 1.6 Illustration of the internet of things encompassing battery-powered devices and everyday connected objects.

1.3.2   The Internet of Everyday Objects

This segment of the internet encompasses applications like retail apparel and footwear, logistics, conveyances, pharmaceutical and healthcare, air baggage and cargo, asset management, inventory and tracking, retail and wholesale, supply chain management, and customer order and delivery tracking. In these applications, electronic identification and sensing devices are attached to low-cost non-electronic everyday objects and wireless gateways are used to connect them to the internet cloud. In the future, non-electronic everyday connected objects are expected to amount to tens or even hundreds of billions and largely outnumber powered devices.

This internet segment requires pervasive wireless identification and sensing technologies of low complexity, low cost, reduced footprint, and low power consumption (ideally battery-free). For example, retail item-level tagging can only be effective if the electronic radio identification device costs a small fraction of the tagged item. On the other hand, pervasive environmental sensing in remote locations may not tolerate frequent battery maintenance. Conventional radio technologies like those mentioned previously are prohibitive in this segment. Here, the adoption of techniques like backscatter, wireless power transfer, energy harvesting, and RFID-enabled sensing is vital [40–44].

Passive UHF RFID technology stands as a strong candidate for pervasive identification and sensing on the internet of things. Recent progress made in UHF CMOS transponder technology, specifically the high level of digital and RF integration achieved, has enabled a dramatic reduction in the price of transponders, making them affordable for item-level tagging. The current number of deployed RFID devices clearly evidences this progress. While there were less than 500 million RFID units sold in 2009, this number amounted to almost 4 billion in 2015 and was projected to be 20 billion in 2020 [45].

Currently, there is a growing effort in the research community to expand the passive RFID concept beyond mere identification to include environmental sensing, data logging, and advanced processing capabilities [46–49]. These concepts are of utmost interest for the internet of things and can open a range of new possibilities.

References

[1] R. G. Carter, Electromagnetism for Electronic Engineers, 3rd ed. Richard G. Carter & Ventus Publishing ApS, Copenhagen, 2010.

[2] A. A. Huurdeman, The Worldwide History of Telecommunications. John Wiley & Sons, Inc., Hoboken, NJ, 2003.

[3] J. C. Maxwell, "A Dynamical Theory of the Electromagnetic Field," *Philosophical Transactions of the Royal Society of London*, 155:459–512, 1865.

[4] E. Fenn, "The Transmitter," EBU Technical Review 263. European Broadcasting Union, Geneva, 1995.

[5] N. Tesla, "The Transmission of Electric Energy Without Wires," *Electrical World and Engineer*, March 5, 1904.

[6] N. Tesla, "Apparatus for Transmitting Electrical Energy," U.S. patent number 1119732, issued December 1914.

[7] W. C. Brown, "The History of Power Transmission by Radio Waves," *IEEE Transactions on Microwave Theory and Techniques*, 32:1230–1242, 1984.

[8] Z. Popovic et al., "Lunar Wireless Power Transfer Feasibility Study," Technical Report DOE/NV/25946-488, 2008.

[9] S. S. Mohammed, K. Ramasamy, and T. Shanmuganantham, "Wireless Power Transmission – A Next Generation Power Transmission System," *International Journal of Computer Applications*, 1(13):100–103, 2010.

[10] N. Shinohara, "Power Without Wires," *IEEE Microwave Magazine*, 12(7):S64–S73, 2011.

[11] J. Glenn, The Complete Patents of Nikola Tesla. Barnes and Noble Books, New York, 1994, pp. 346–360.

[12] W. C. Brown, "The History of the Development of the Rectenna," in *Proc. SPS Microwave Systems Workshop at JSC-NASA*, 1980, pp. 271–280.

[13] W. C. Brown, J. R. Mims, and N. I. Heenan, "An Experimental Microwave-Powered Helicopter," *IEEE International Convention Record*, 13(5):225–235, 1965.

[14] P. E. Glaser, "Power from the Sun; Its Future," *Science*, 162(3856):857–886, 1968.

[15] US Department of Energy / NASA. *Satellite Power System, Concept Development and Evaluation Program: Reference System Report*, Technical Report DOE/ER-0023. US Department of Energy, 1979.

[16] H. Matsumoto, "Research on Solar Power Station and Microwave Power Transmission in Japan," *IEEE Microwave Magazine*, 3(4):36–45, 2002.

[17] J. J. Schelesak, A. Alden, and T. Ohno, "A Microwave Powered High Altitude Platform," in *IEEE MTT-S International Microwave Symposium Digest*, 1988, pp. 283–286.

[18] A. Kurs, et al., "Wireless Power Transfer via Strongly Coupled Magnetic Resonances," *Science*, 317(5834):83–86, 2007.

[19] C.-C. Lo, et al., "Novel Wireless Impulsive Power Transmission to Enhance the Conversion Efficiency for Low Input Power," in *IEEE MTT-S International Microwave Workshop Series on Innovative Wireless Power Transmission*, 2011, pp. 55–58.

[20] Wireless Power Consortium, "Qi WPT Standard: System Description Wireless Power Transfer, Volume I: Low Power, Part 1: Interface Definition, Version 1.1.2," 2013. Available at: www.wirelesspowerconsortium.com/downloads/wireless-power-specification-part-1.html

[21] Powercast, "Wireless Power Innovations." [Online], available at: www.powercastco.com

[22] Wikipedia, "eCoupled." [Online], available at: https://en.wikipedia.org/wiki/ECoupled

[23] Wikipedia, "WiPower, The Alliance For Wireless Power." [Online], available at: https://en.wikipedia.org/wiki/WiPower

[24] Powermat, "Wireless Power Solutions." [Online], available at: http://powermat.com

[25] J. L. Baird, "Improvements in or Relating to Apparatus for Transmitting Views or Images to a Distance," patent number GB292185, 1928.

[26] S. A. Weis, "RFID (Radio Frequency Identification): Principles and Applications," *System*, 2(3):1–23, 2007.

[27] H. Davis, "Eavesdropping Using Microwaves – Addendum," *EE Times*. [Online], available at: www.eetimes.com/eavesdropping-using-microwaves-addendum

[28] P. V. Nikitin, "Leon Theremin (Lev Termen)," *IEEE Antennas and Propagation Magazine*, 54(5):252–257, 2012.

[29] G. Brooker and J. Gomez, "Lev Termen's Great Seal Bug Analyzed," *IEEE Aerospace and Electronic Systems Magazine*, 28(11): 4–11, 2013.

[30] H. Stockman, "Communication by Means of Reflected Power," *Proceedings of the Institute of Radio Engineers*, 36(10):1196–1204, 1948.

[31] D. B. Harris, "Radio Transmission Systems with Modulatable Passive Responder," U.S. patent number 2927321, 1960.

[32] K. Finkenzeller, "*RFID Handbook*," 2nd ed. Wiley, Chichester.

[33] M. W. Cardullo, "Transponder Apparatus and System," U.S. patent number 3713148A, 1973.

[34] J. I. Aguirre, "EPCglobal: A Universal Standard," MSc thesis, Massachusetts Institute of Technology, 2007.

[35] EPCglobal Inc., "EPC Class-1 Generation-2 UHF RFID, Protocol for Communications at 860 MHz–960 MHz, Version 1.2.0."

[36] H. Yue, et al., "DataClouds: Enabling Community-Based Data-Centric Services Over the Internet of Things," *IEEE Internet of Things Journal*, 1(5):472–482, 2014.

[37] Ericsson, "More than 50 Billion Connected Devices," Ericsson White Paper 284 23-3149 Uen, 2011.

[38] J. A. Stankovic, "Research Directions for the Internet of Things," *IEEE Internet of Things Journal*, 1(1):3–9, 2014.

[39] D. K. McCormick, "Preview," in *IEEE Technology Report on Wake-Up Radio*, 2017, pp. 1–11.

[40] S. Amendola, et al., "RFID Technology for IoT-Based Personal Healthcare in Smart Spaces," *IEEE Internet of Things Journal*, 1(2):144–152, 2014.

[41] M. S. Khan, M. S. Islam, and H. Deng, "Design of a Reconfigurable RFID Sensing Tag as a Generic Sensing Platform Toward the Future Internet of Things," *IEEE Internet of Things Journal*, 1(4):300–310, 2014.

[42] H. Hammad, "New Technique for IoT Indoor Localization by Employing Segmented UHF RFID Bandwith Using Bandpass Filters and Circulators," in *2018 18th International Symposium on Antenna Technology and Applied Electromagnetics*, 2018, pp. 1–4.

[43] B. S. Çiftler, A. Kadri, and İ. Güvenc, "IoT Localization for Bistatic Passive UHF RFID Systems With 3-D Radiation Pattern," *IEEE Internet of Things Journal*, 4(4):905–916, 2017.

[44] I. Farris, et al., "6lo-RFID: A Framework for Full Integration of Smart UHF RFID Tags into the Internet of Things," *IEEE Network*, 31(5):66–73, 2017.

[45] M. Hetrick, "The Role of RFID in the Identification of Things," ProMat, Chicago, 2017.

[46] G. Zhang et al., "Architecture Characteristics and Technical Trends of UHF RFID Temperature Sensor Chip," *Active and Passive Electronic Components*, 2018:9343241, 2018.

[47] J. Gao, et al, "Printed Humidity Sensor with Memory Functionality for Passive RFID Tags," *IEEE Sensors Journal*, 13(5):1824–1834, 2013.

[48] D. J. Yeager, A. P. Sample, and J. R. Smith, "WISP: A Passively Powered UHF RFID Tag with Sensing and Computation," in *RFID Handbook: Applications, Technology, Security, and Privacy,*" S. A. Ahson and M. Ilyas, CRC Press, Boca Raton, 2008.

[49] S. Kim, et al., "Low-Cost Inkjet-Printed Fully Passive RFID Tags for Calibration-Free Capacitive/Haptic Sensor Applications," *IEEE Sensors Journal*, 15(6):3135–3145, 2015.

2   An Introduction to Radio Frequency Identification

RFID has become an undeniable part of our personal and business lives, with applications spanning from access control and logistics to healthcare and aircraft maintenance. But with the recent development of the internet of things, RFID technology has gained even more importance. The ubiquitous connection of every-day objects anywhere, anytime in the IoT requires low-cost, low-complexity, and low-power hardware, and RFID and its underlying concepts like backscatter communication and wireless power transmission have the potential to meet these requirements. With a total market size of 10 billion USD in 2017, the total RFID market (including transponders, readers, software, and services) is expected to be worth over 13 billion in 2022 [1]. This fast-growing field will certainly continue to present new opportunities for engineers and researchers. In this chapter, we introduce the key concepts, applications, standards, and architectures of modern RFID technology.

2.1   Basic Concepts and Definitions

Radio frequency identification is an automatic wireless data capture technique used to identify objects by means of electronic devices called transponders or tags. The transponder typically comprises a memory containing a unique serial number associated to the object to be identified, and a radio-frequency circuit and antenna to communicate wirelessly. A radio device called an interrogator or reader is used to wirelessly access the transponder memory. Figure 2.1 illustrates a basic RFID system including RFID transponder, reader, and antennas.

RFID systems can be categorized according to the method used to power and communicate with the transponder (Section 2.1.1), transponder memory capacity (Section 2.1.2), the coupling between the reader and transponder (Section 2.1.3), system architecture, and so on. Depending on the type of power supply and communication method, transponders can be passive, semi-passive, or active. Passive transponders present minimal cost and complexity as they do not require onboard power sources, local oscillators, or other complex components. This makes them ideal for high-volume item-level identification.

Regarding the coupling between the reader and transponder, RFID systems fall into two main categories: long-range electromagnetically coupled systems typically

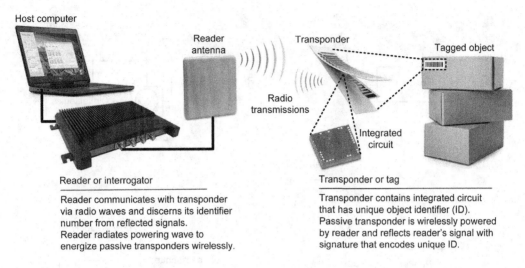

Reader or interrogator

Reader communicates with transponder via radio waves and discerns its identifier number from reflected signals. Reader radiates powering wave to energize passive transponders wirelessly.

Transponder or tag

Transponder contains integrated circuit that has unique object identifier (ID). Passive transponder is wirelessly powered by reader and reflects reader's signal with signature that encodes unique ID.

Figure 2.1 Basic illustration of UHF radio frequency identification system.

operating in the ultra-high frequency (UHF) and microwave bands [2], and short-range inductively coupled systems typically operating in low-frequency (LF) and high-frequency (HF) bands [3]. Proximity smart cards operating in the 13.56 MHz band are the most widespread application of the inductive coupling technique, whereas passive UHF inlays are the most popular application of the electromagnetic coupling method.

The development of passive UHF RFID has traditionally been driven by the need to replace optical barcodes with an enhanced identification method, and RFID can offer a faster, longer-range, and more robust alternative. Despite these potential benefits, cost has traditionally been a major roadblock to the adoption of RFID as a replacement for barcodes. But recent progress made in CMOS transponder technology has allowed dramatic reductions in the cost of UHF RFID inlays and transponders, and has taken the technology closer to mainstream adoption.

Although RFID is traditionally used for mere identification, there has been an effort to expand the concept of passive transponder beyond identification to feature sensing, data logging, and advanced processing capabilities. This emerging concept, known as passive RFID-enabled sensor or passive wireless sensor [4, 5], has gained traction in the internet of things where the energy autonomy of wireless nodes is vital.

2.1.1 Passive, Active, and Semi-Passive Systems

Depending on their power supply method, RFID transponders can be classified as passive or active. Active transponders use onboard power supplies / batteries and standard RF components found in conventional wireless radio transceivers such as a local oscillator, transmitter, and receiver (see Figure 2.2(a)). These kinds of transponders can usually communicate over long distances and are limited primarily by their transmit power and receive sensitivity. Active transponders usually have large

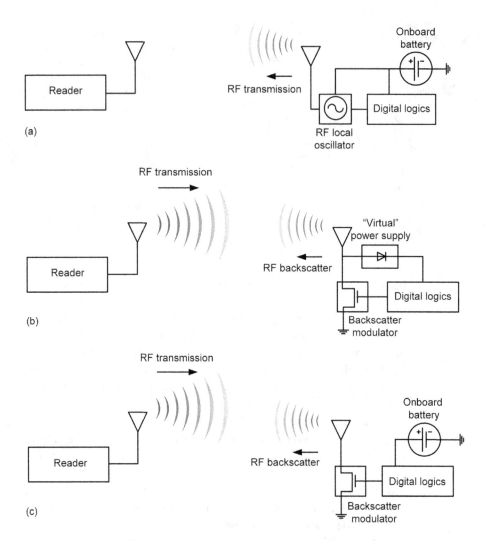

Figure 2.2 Illustration of RFID transponder power supply and communication methods. (a) Active RFID system. (b) Passive RFID system. (c) Semi-passive RFID system. Adapted from [2].

memory capacity, can perform relatively complex computations, and do certain tasks "offline." For example, an active transponder could collect data from a temperature sensor offline, store that data in memory, and transfer it to the network when interrogated by a reader. The downsides of active transponders, which make them unsuitable for high-volume item-level tagging, include increased complexity, cost, size, and maintenance.

Passive transponders require no batteries and instead harvest energy to power up their electronics from the reader magnetic or electromagnetic field (see Figure 2.2(b)). These kinds of transponders have no local oscillator and reflect the signal received from the reader with a given signature to communicate their message. In electromagnetically

Table 2.1 Passive, semi-passive and active transponders.

	Passive	Semi-passive	Active
Onboard power supply	No	Yes	Yes
Onboard RF oscillator	No	Typically, no	Yes
Communication method	Energy reflection	Energy reflection	Self-generated signal
Coverage range	Short (up to 10 m)	Medium (up to 30 m)	Long (> 100 m)
Cost and complexity	Low	Medium	High
Type of memory	Mostly read-only	Read-write	Read-write

coupled systems, this method is known as backscatter modulation and consists of modulating the radar cross section of the transponder antenna (see Figure 2.2(b) and (c)). In inductively coupled systems, transponders modulate the mutual coupling between the reader and transponder coil antennas. Compared to their active counterparts, passive transponders are simpler, smaller, cheaper, and require no maintenance. But they present shorter communication ranges and are computationally limited.

Semi-passive or semi-active (sometimes referred to as battery-assisted) transponders employ batteries to power up their digital electronics as in active transponders, but do not require a local oscillator and instead reflect the reader signal to convey their message just like passive transponders. These systems present a tradeoff offering longer communication distance than passive systems and longer battery lifetime than active systems. Table 2.1 relates the main features of the different types of transponders.

2.1.2   One-Bit versus *N*-Bit Transponder Systems

Regarding their information capacity, RFID transponders can be classified as 1-bit or *N*-bit. One-bit systems can only inform about the presence or absence of transponders in the field of the reader and are typically used in counter-theft applications in supermarkets and retail and apparel stores. *N*-bit transponders, on the other hand, can store identifiers with up to 100 or more bits that enable unique identification of millions or billions of objects. For example, the 96-bit EPC system can accommodate up to 268 million unique company IDs (28 bits), each with 16 million object classes (24 bits) and 68 billion serial numbers per class (36 bits). This is more than enough to cover all products manufactured worldwide for years to come. Modern *N*-bit transponders can also offer advanced features like writable memory space for storing user data.

2.1.3   Near-Field and Far-Field Coupling

The main coupling mechanisms used in modern RFID involve near-field inductive coupling and far-field electromagnetic coupling. The former mechanism is used for short-range applications at low and high frequencies (typically at 13.56 MHz and below 135 kHz), and the latter mechanism is used for long-range applications at UHF and microwave frequencies (typically, 433 MHz, 860–960 MHz, 2.45 GHz, and 5.8 GHz).

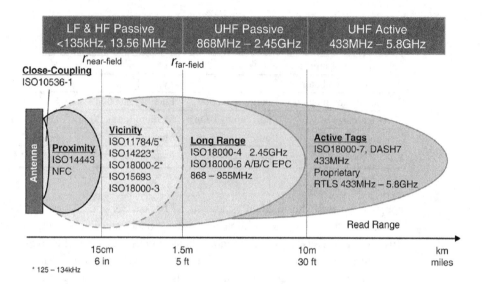

Figure 2.3 Illustration of antenna near- and far-field operating regions. Courtesy of Texas Instruments [7].

Figure 2.3 illustrates typical frequency bands, methods, and communication ranges for each type of coupling. The near-field and far-field limits ($r_{\text{near-field}}$, $r_{\text{far-field}}$) can be estimated from the signal wavelength and antenna geometry (see Chapter 2 of [6]).

2.2   RFID Applications

Here, we discuss the most emblematic traditional applications of RFID technology including retail supply chain, healthcare, logistics, access control, and biometric identification. We present specific examples and highlight key benefits offered by the technology in these segments.

2.2.1   Retail Supply Chain

The retail supply chain has traditionally been the main target segment for the application of UHF RFID [2, 8]. The potential benefits of RFID in this segment include streamlining of business processes, improved inventory visibility, more efficient and faster shipping, and savings in operational and labor costs. The mandate from Walmart to its suppliers to tag their pallets and cases [2] and the adoption of RFID by the U.K.'s Tesco Corporation for tagging of cases in distribution centers [9] are two of the most historically relevant examples in this segment. Other major retailers that have used item-level RFID tagging to improve operations and consumer experience include Target, Macy's, and Levi Strauss [10].

In the supply chain, RFID can help organizations overcome the main difficulties faced in the process of delivering goods such as incorrect shipment, late delivery, difficulty locating goods, difficulty reconciling customer orders and returns, misplaced goods, and inaccuracies in forecasts. Besides the obvious gains for the retail supply chain segment in terms of inventory visibility and operational cost savings, RFID also offers a pathway to thwarting counterfeiting, theft, and product diversion.

2.2.2   Pharmaceutics and Healthcare

RFID technology has proven very attractive to the pharmaceutical and healthcare sectors [11–14]. The technology has been used in hospitals to efficiently identify patients and label and track assets with potentially fewer human errors and reduced costs [11, 12]. Future applications in this sector promise to go beyond logistics – for example, RFID-enabled health sensors embedded in body prostheses can allow healthcare professionals to monitor the improvement of their patients [13]. The potential of these approaches for the healthcare sector can be increased significantly when combining RFID with modern mobile computing [14]. In the pharmaceutical segment, RFID tagging can also help to combat counterfeit medicines [12].

2.2.3   Aircraft Maintenance and Airline Operations

In the airline industry, RFID is potentially beneficial for applications like baggage handling, catering, cargo handling, passenger ticketing, employee badging, security processes, and management of critical aircraft parts. Airbus and Boeing Aerospace, the world's two major commercial airplane manufacturers, have successfully deployed RFID to improve efficiency and efficacy in their processes. For example, Airbus used RFID for the identification, management, and traceability of critical airplane parts through manufacturing and maintenance [15], and Boeing employed RFID to manage emergency equipment like life vests, oxygen generators, and other essential cabin items [16].

Baggage handling is perhaps the most prominent area of application of RFID in the airline sector. Given the proven benefits of RFID for baggage identification and tracking, which include reduction of operational costs and baggage mishandling, the airline industry has embraced RFID on a global scale [17]. After several studies and recommendations, this global adoption has been consolidated with the International Air Transport Association (IATA) unanimously voting in their 2019 annual general meeting to support the global deployment of RFID for airline baggage tracking [18]. IATA's resolution also considers the implementation of baggage messaging standards for more accurate tracking of baggage in real time at key points of their itinerary.

2.2.4   Biometric Identification

Contactless RFID has been used to improve the level of security of digital or biometric passports [19]. Storing the passport holder's biometric information (digital imaging and fingerprint scan) in an RFID chip embedded in the passport provides an unparalleled level of security, making it harder to forge the document. Currently, more than 100 countries issue RFID-based biometric passports.

2.2.5   Other Applications

Other popular RFID applications include access control, library management, animal identification, passive keyless entry, electronic toll collection, container identification, railcar identification, and electronic banking and payment.

2.3   RFID Standards and Regulations

Standards and regulations play a key role in the global adoption of RFID. An RFID standard or protocol establishes the set of rules and guidelines of the "language" to be used in a conversation between compliant RFID readers and transponders to ensure interoperability between systems from different manufacturers. A standard defines, for example, the bit rate, timing, modulation, and encoding schemes to be used in the communication between a reader and transponder.

Regulations enforced by local agencies lay down the laws and rules for efficient and safe use of technology. Regulations define, for example, radio channels and maximum radiated power levels. Radio spectrum regulations have a twofold purpose – ensure that the spectrum is shared by multiple users in an effective and efficient manner, and guarantee safe operation for humans and animals. In this section, we give an overview of some aspects of UHF RFID standardization and regulation. In Chapter 4, we discuss the ISO 18000-63 standard in more detail. For further information on RFID standards and regulations, the interested reader is referred to [2] and [3].

2.3.1   RFID Regulations

Radio frequency identification regulations ensure that electromagnetic radiation from RFID systems do not disrupt the operation of other radio systems and is safe for human exposure. Regulations typically vary from country to country and are established by various agencies including the International Telecommunication Union (ITU), US Federal Communications Commission (FCC), European Telecommunications Standards Institute (ETSI), and European Radiocommunications Committee (ERC). The ITU deals with technical and administrative aspects of telecommunications and is responsible for the standardization, allocation, and coordination of the radio spectrum. In Europe, the ERC provides recommendations that serve as a basis for legislation and licensing of radio systems at the national level. The main aspects of regulations include:

- channel allocation and channel spacing
- duty cycle of the transmitted frame
- effective isotropic radiated power (EIRP)
- signal bandwidth
- transmit spectral mask.

Sharing the radio spectrum among several users effectively and efficiently is challenging – the radio spectrum is limited, and as the number of deployed radio systems increases,

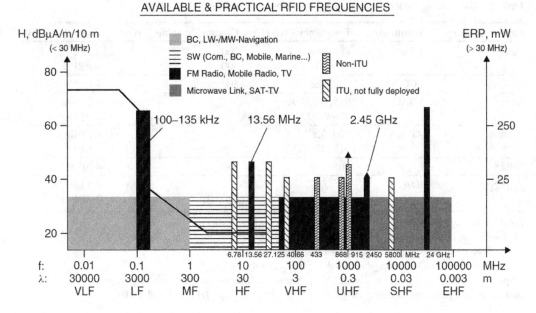

Figure 2.4 Radio spectrum allocation. Reprinted from [3].

the spectrum becomes more congested and greater efficiency is required in the way it is used.

On the other hand, finding common free spectrum slots across multiple regions for new systems is a difficult task. For example, while UHF RFID systems can operate in the 902–928 MHz band in the US, this is not possible in Europe because that portion of the spectrum has already been allocated for other applications. For operation in multiple regions, multi-band RFID designs are typically required. The EPC Gen2 and ISO 18000-63 standards enable interoperability worldwide in the UHF band. Figure 2.4 presents the overall spectrum allocation worldwide and Table 2.2 summarizes the main aspects of UHF RFID regulations in several regions. Refer to [20] for further details and spectrum allocations in other regions.

2.3.2 Regulation in the US – FCC Part 15

In the US, RFID devices can operate in the unlicensed industrial, scientific, and medical (ISM) bands 902–928 MHz, 2400.0–2483.5 MHz, and 5725–5850 MHz under conditions specified in the FCC Part 15 rules section 15.247 [21]. This regulation is part of Title 47 of the US Code of Federal Regulations (CFR) that covers electromagnetic compliance and is regulated by the FCC. RFID systems compliant with FCC Part 15 use frequency-hopping spread-spectrum modulation and UHF readers can radiate a maximum power of 1 W, or up to 4 W with a directional antenna if they randomly hop among a minimum of 50 channels. For the default 500 kHz channel spacing, RFID interrogators can use the spectrum from 902.75 MHz to 927.25

Table 2.2 UHF RFID regulators, and reference frequencies and powers in different regions.

Region/Country	Regulator	Frequency [MHz]	Channel spacing [kHz]	Power limit [W]
Europe*	ETSI	865–868	200	2 ERP
		915–921	400	4 ERP
USA	FCC	902–928	500	4 EIRP
Japan	MIAC	916.7–920.9	—	4 EIRP
		916.7–923.5		0.5 EIRP
Australia	ACMA	920–926	—	4 EIRP
		918–926		1 EIRP
India	DOT	865–867	200	4 EIRP
Singapore	IDA	866–869		0.5 ERP
		920–925		2 ERP

MIAC – Japanese Ministry of Internal Affairs and Communications
DOT – Indian Department of Telecommunications
IDA – Infocomm Development Authority of Singapore
ACMA – Australian Communications and Media Authority
* Specific regional restrictions apply (see [20]). Currently, interrogator transmissions in the 4 W ERP upper
  band are only permitted at center frequencies 916.3 MHz, 917.5 MHz, and 918.7 MHz.

MHz, covering a frequency range of 24.5 MHz allowing for a 750 kHz guard band at
the band edges. For further details about RFID regulations in the US, refer to [2].

2.3.3   Regulation in Europe – ETSI EN 302 208

The first standard to regulate RFID use in Europe was ETSI EN 300-220 [22]. This
regulation broadly applies to short-range devices (SRD) operating in the 25–1000
MHz frequency range, including RFID devices. The operation of UHF RFID in
Europe specified within this regulation was initially limited to a single 250 kHz
channel from 869.4 MHz to 869.65 MHz at a maximum transmitted power of 0.5
W ERP. This restrictive regulation was later ratified by the ETSI EN 302 208 standard,
which introduced key improvements, principally an extended frequency band from
865 MHz to 868 MHz, shared RFID operation in 15 channels of 200 kHz, transmit
power levels up to 2 W ERP, and mandatory "listen before talk" capability [23]. These
improvements enabled UHF RFID reader performance in the EU comparable to those
achieved under FCC rules in the US.

ETSI EN 302 208 later introduced an additional band from 915 MHz to 921 MHz
with power levels up to 4 W [24]. The frequency usage conditions for RFID operation
in the EU are currently harmonized in the lower band (865–868 MHz) according to
[25] and in the upper band (915–921 MHz) according to [26]. Figure 2.5 shows the
channel plan of the lower band according to ETSI 302 208. The upper band, which
only permits three channels, currently has a limited implementation status within the
EU and participating CEPT (European Conference of Postal and Telecommunications
Administrations) countries [27].

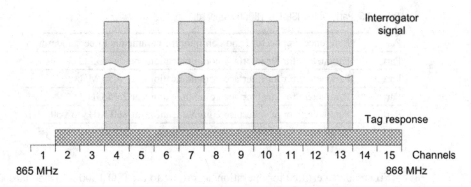

Figure 2.5 Channel plan for the lower UHF band under ETSI 302 208. Reprinted from [24].

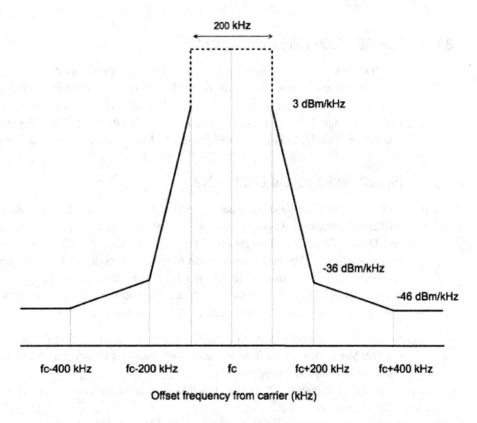

Figure 2.6 Spectrum mask for the lower UHF band under ETSI 302 208. Reprinted from [24].

Local RFID regulations also define transmit spectrum masks to limit excessive radiation outside the allowed bandwidth and minimize out-of-channel and out-of-band interference. Figure 2.6 depicts the spectrum mask requirement for systems operating under the ETSI 302 208 regulation. In addition to meeting local regulations, UHF

Table 2.3 Parts of the ISO/IEC 18000 standard.

Part 1	Reference architecture and definition of parameters to be standardized
Part 2	Parameters for air interface communications below 135 kHz
Part 3	Parameters for air interface communications at 13.56 MHz
Part 4	Parameters for air interface communications at 2.45 GHz
Part 6	Parameters for air interface communications at 860 MHz to 960 MHz
Part 7	Parameters for active air interface communications at 433 MHz

RFID readers certified for operation according to the ISO 18000-63 standard shall also meet spectrum masks defined in that standard. Chapter 5 discusses filtering and coding/modulation schemes for efficient spectrum efficiency and transmit mask compliance.

2.3.4   The ISO 18000 Family of Standards

ISO/IEC 18000 covers an international family of standards, developed and maintained by the joint technical committee of the ISO and the International Electrotechnical Commission (IEC), that describes various RFID technologies, each using a unique frequency range. Table 2.3 presents the various parts of ISO/IEC 18000 under the general title *Information Technology – Radio Frequency Identification for Item Management*.

2.3.5   ISO/IEC 18000-63 and EPC Gen2V2

Part 6 of ISO/IEC 18000 describes the parameters for air interface communications utilizing electromagnetic propagation at 860–960 MHz and comprises Types A, B, C, and D (see Table 2.4). The ISO/IEC 18000-63 standard, which defines communications of Type C, resulted from an effort to replace barcodes with a more robust and reliable identification method based on RFID. This standard has become the most widely used RFID standard in the UHF band. The ISO/IEC 18000–63 evolution since the barcode technique was introduced in 1973 is summarized below.

- 1973: GS1's barcode is the first single standard for product identification.
- 1999: Massachusetts Institute of Technology Auto-ID Center founded to develop the electronic product code (EPC).
- 2003: GS1 launches EPCglobal to facilitate the development and adoption of EPC/ RFID standards.
- 2004: GS1 publishes the first version of the EPC Gen2 Air Interface standard.
- 2005: ISO/IEC incorporates the EPC Gen2 standard into ISO/IEC 18000-6C.
- 2008: GS1 releases new EPC version to improve RFID performance for item-level tagging.
- 2009: GS1 releases guidelines for EPC-based Electronic Article Surveillance (EAS).
- 2010: Industry working group launched to enhance the Gen2 standard for UHF, based on requests for additional functionalities from EPC user community.

Table 2.4 Summary of air interface characteristics of ISO/IEC 18000 Part 6.

	ISO 18000-61 (Type A)	ISO 18000-62 (Type B)	ISO 18000-63 (Type C)	ISO 18000-64 (Type D)
Release date (first version)	2004	2004	2008	2010
Frequency band*	860–960 MHz			
Reader encoding	PIE	Manchester	PIE	
Reader bit rate	33 kbps	10 or 40 kbps	up to 128 kbps	
Reader modulation	ASK	ASK	ASK/PR-ASK	
Transponder encoding	FM0	FM0	FM0 or Miller	Pulse position encoding / Miller
Transponder bit rate	40 or 160 kbps	40 or 160 kbps	up to 640 kbps	
Transponder modulation	ASK	ASK	ASK or PSK	
Anti-collision algorithm	ALOHA	Binary tree	Random slotted (Q-Algorithm)	Tag only talks after listening (TOTAL)

* The specific frequency band is region-dependent (see Table 2.2).

- 2013: First major update since 2008.
- 2014: ISO incorporates EPC Gen2v2 into the ISO/IEC 18000-63 standard.

Although EPC Gen2v2 and ISO/IEC 18000-63 are not necessarily the same, they share the same air interface parameters and are used interchangeably throughout this book.

2.3.6   Other RFID Standards

ISO and IEC maintain several inductive contactless standards for close-coupling smart cards (e.g., ISO/IEC 10536), proximity-coupling smart cards (e.g., ISO/IEC 18000-3, ISO/IEC 14443 Types A and B, and ISO/IEC 18092) and vicinity-coupling smart cards (e.g., ISO/IEC 15693). Other popular inductive contactless standards include MIFARE from NXP, FeliCa from Sony, and Near Field Communication (NFC) maintained by the NFC forum.[1] The ISO and IEEE also maintain standards for long-range active RFID and sensor applications, including ISO/IEC 18000-7 and IEEE 802.15.4f.

---

[1] The NFC forum was created by Nokia, Sony, and Philips and was later joined by several other manufacturers, applications developers, and financial services institutions like Apple, Google, Samsung, and Visa.

2.4   RFID Transponders and their Operating Principles

In this section, we describe the most popular types of RFID transponders with the focus on their operating principles. For better understanding, we also include a description of the reader when necessary.

2.4.1   Inductive 1-Bit Transponders

Inductively coupled 1-bit systems are the simplest RFID detection systems. These systems allow the detection of the presence of transponders inductively coupled to the reader coil antenna at a certain resonance frequency. Figure 2.7 illustrates a 1-bit LC transponder system, where a capacitor is connected in parallel with the transponder coil antenna to form a resonant LC circuit tuned to a particular frequency and the reader uses a sweep or chirp transceiver to scan the interrogation field. When the reader swept frequency matches the transponder's resonance frequency, the reader coil antenna gets loaded which causes a voltage drop across its terminals. By sensing this voltage drop, the reader can detect the presence of the transponder in its field.

One-bit transponders can also be made of magnetically sensitive metal strips mechanically resonant at certain frequencies [2]. This kind of transponder, which was first introduced in the 1960s to thwart shoplifting, is still in use today. Inductively coupled 1-bit systems are compact and inexpensive, and since they operate at very low frequencies (tens of kilohertz to a few megahertz) they have a straightforward design and construction.

2.4.2   Inductive N-Bit Transponders

Inductively coupled N-bit transponders collect the energy required to operate their electronics from the reader's magnetic field and load-modulate their coil antennas to transfer information to the reader (Figure 2.8). In these systems, the reader-to-transponder energy transfer happens in the same way as in a voltage transformer, with the

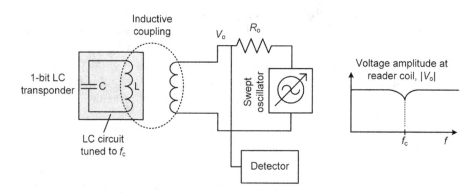

Figure 2.7 Inductively coupled 1-bit RFID system [U.S. Patent 3752960].

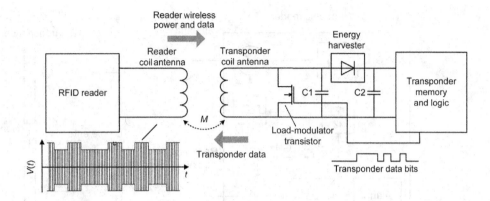

Figure 2.8 Near-field inductively coupled passive RFID system. Adapted from [3].

reader antenna acting as the transformer's primary coil and the transponder antenna acting as the secondary coil (coupled to the primary via a mutual inductance M). By applying an alternating (AC) current to the reader coil antenna, one creates an AC magnetic field that yields an AC voltage on the transponder coil.

The AC energy induced in the transponder coil antenna is converted to DC power by an RF to DC converter circuit, and the collected DC power is used to supply the digital electronics of the transponder. To boost energy transfer efficiency, a frequency-tuned impedance-matching circuit is typically used at each coil antenna [3], and over-voltage protection, voltage regulation, and power management circuitry is used to stabilize the collected DC power.

To communicate information to the reader, the transponder load-modulates its coil antenna according to the information bits to be transmitted. This is typically done by modulating the gate of a transistor which is connected across the transponder coil antenna (see Figure 2.8). Switching this transistor on and off modulates the load impedance presented to the transponder coil antenna which, in turn, induces changes across the reader coil antenna. The reader retrieves the transponder information by demodulating and decoding the voltage changes induced on its coil antenna. Refer to [3] for more information on the operation and design of inductive systems.

The most popular N-bit transponder systems using inductive coupling are proximity smart cards [28] and NFC devices [29] with coverage ranges of several centimeters, and vicinity smart cards [30] with coverage ranges up to 1 meter. These systems typically operate at LF (100–135 kHz) and HF (10–15 MHz) bands.

2.4.3 Harmonic 1-Bit Transponders

Harmonic 1-bit transponder systems typically use electromagnetic propagation in the microwave range and exploit the harmonic distortion mechanism of a non-linear device, typically a radio-frequency diode. The right side of Figure 2.9 shows a harmonic 1-bit transponder consisting of an RF diode attached to a dipole antenna.

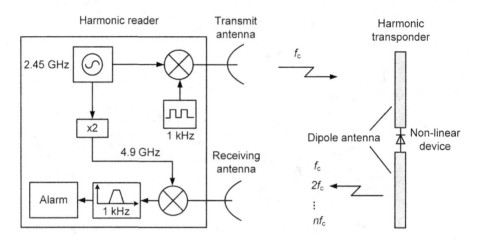

Figure 2.9 Microwave harmonic 1-bit system. Adapted from [3].

When the transponder is illuminated with a microwave interrogation signal (e.g., at 2.45 GHz), it produces and re-radiates harmonics of that signal (at 4.9 GHz, 7.35 GHz, 9.8 GHz, . . .).

To detect transponders in the interrogation field, the reader (left side of Figure 2.9) examines one of the harmonics radiated by the transponder (usually the second harmonic) by homodyning the received signal using a local oscillator signal at the harmonic frequency. To enhance detectability, the interrogation signal in the transmitter is commonly modulated using a subcarrier, and the homodyned signal is bandpass filtered to eliminate out-of-channel interference (see Figure 2.9). Other advanced approaches based on non-linear intermodulation distortion have also been proposed [31].

2.4.4   Backscatter *N*-Bit Transponders

Figure 2.10 illustrates an electromagnetically coupled *N*-bit passive-backscatter system. The reader broadcasts an unmodulated carrier that is used to wirelessly power the transponder and serves as a remote carrier for the transponder-to-reader communication. By reflecting a portion of the carrier signal received from the reader with a specific signature, the transponder can communicate information to the reader without having to locally generate a radio signal. Conceptually, this procedure looks like that used in inductive *N*-bit systems, with the exception that here the reader-to-transponder coupling is achieved via electromagnetic propagation. Like in inductive systems, a transistor is used to modulate the antenna (see Figure 2.10).

The simplest backscatter system uses two modulation states, absorptive and reflective. In the absorptive state, the modulating transistor is turned off and the impedance of the RFID chip is matched to the transponder antenna to minimize the power the transponder reflects to the medium while maximizing the power it collects from the incoming signal. The reflective state is achieved by turning on the transistor and short-

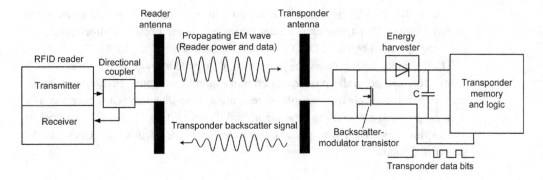

Figure 2.10 Passive-backscatter system. Adapted from [3].

circuiting the transponder antenna. Commercial UHF RFID transponders typically produce binary amplitude-shift keying (ASK) modulation by alternating the termination presented to the antenna between a matched load that promotes maximum power transference to the chip and a mismatched resistive load that increases the amplitude of the reflected signal without altering its phase. Binary phase-shift keying (PSK) modulation can be produced by switching a capacitor across the transponder antenna to create a phase shift [2, 32].

Highly reflective states can impair transponder operation due to low power availability to the RFID chip. On the other hand, minimally reflective states penalize the backscatter radio communication link. Therefore, a tradeoff between backscattered signal and power available to the chip is required in the design of backscatter transponders [2]. As UHF RFID transponders are typically forward-link-limited (see Chapter 3), it may be useful to trade backscattered signal for power available to the chip.

Backscatter readers typically share the same local oscillator for transmitting and receiving and can use two separate antennas (bistatic configuration) or a shared antenna (monostatic configuration) to transmit and receive. The latter configuration, illustrated in Figure 2.10, requires a circulator or directional coupler to separate the transponder-backscattered signal from the reader-transmitted carrier. In both configurations, the total signal received by the reader is a combination of transponder-backscattered signals plus undesired, unavoidable signal components due to clutter reflections and transmitter-to-receiver leakage.

Since these unwanted signal components can be much larger than the backscatter signal of interest and have unpredictable amplitudes and phases, the effect of the often-weak transponder modulation on the total received signal is unpredictable and uncontrollable. The only certainty is that a change in the transponder modulation state changes something about the phase and/or amplitude of the received signal. For an effective backscatter link, the transponder information must be encoded on these changes and not on the amplitude, phase, or direction of the modulation states [2]. Practical transponder encoding schemes are usually variations of frequency-shift keying (FSK) and rely on counting and timing the number of state transitions in the received signal. This approach is used, for example, in the FM0 and Miller subcarrier encoding specified by the ISO/IEC 18000-63 standard (see Chapter 4).

The example above considers only two transponder modulation states, but a transponder can generally produce an arbitrary number of states. Higher-order back-scatter modulation schemes capable of generating more than two states have recently been proposed to increase the data rate in backscatter systems. We will have more to say on this topic when we discuss quadrature amplitude modulation (QAM) back-scatter in Chapter 3.

2.4.5   RFID-Enabled Sensors

Passive-backscatter transponders are unique wireless communication systems in that they require no local power source, radio-frequency oscillator, or other complex components like mixers or amplifiers. They present, therefore, minimal power consumption, complexity, size, and cost, which makes them ideal for pervasive item-level tagging. In addition to identification functions, ideal passive-backscatter transponders would feature environmental sensing, data logging, and advanced computation capabilities at minimal complexity and negligible power consumption.

While some of these requirements are conflicting, significant progress has been made toward expanding the passive transponder concept beyond mere identification to include functionalities like sensing. This concept is known as a passive RFID-enabled sensor, passive wireless sensor, or wirelessly powered sensor [4, 5]. An example of a commercial RFID device offering sensing capabilities is the SL900A transponder [33], which incorporates a temperature sensor and an analog-to-digital converter (ADC) interface for sampling external sensors. This chip also features a digital serial peripheral interface (SPI) for communicating with external devices. Figure 2.11 shows a generic architecture for a passive UHF RFID transponder with sensing capability comprising the following building blocks [4, 34, 35]:

1. Energy harvesting circuit – key part of a passive transponder, responsible for converting the radio-frequency power collected by the transponder antenna into DC power to supply the transponder electronics. Voltage multiplier or charge pump circuits are typically used to boost the rectified voltage to a level high enough to supply the digital electronics of the transponder. To promote maximum power transfer to the transponder chip, the antenna impedance should be conjugate-matched to the input impedance of the chip. Transponder impedance matching is typically performed at the high-impedance (absorptive) state of the chip in which the backscatter modulation transistor is off (see Figure 2.11).

2. Power management unit – follows the energy harvesting circuit and typically comprises voltage limiting, energy storage, voltage regulation, voltage supervision, and power-on-reset circuitry to deliver a continuous, regulated DC power supply to the transponder electronics. The power management unit counteracts power supply fluctuations caused by variations in the load current due to transients in the

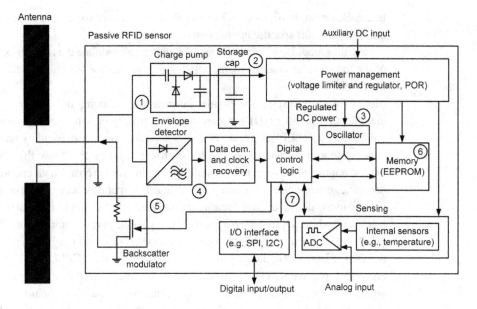

Figure 2.11 Architecture of a passive UHF RFID transponder with sensing capabilities.

transponder digital circuitry, varying operating distance, and changes in the radio channel, including momentaneous drop or interruption of the reader-powering signal.

3. Clock oscillator – provides the clock required for operating the digital control logic and timing the baseband data. To minimize transponder consumption, the oscillator usually runs at a low frequency. In some implementations, the clock required for transponder operation is encoded in the reader signal, eliminating the need for a local clock generator in the transponder [4]. This can significantly lower the power consumption requirement in the transponder while reducing its complexity and enabling scalability to various data rates for different power and distance constraints [35].

4. Demodulator – encompasses envelope detection, data demodulation, and clock recovery functions. The modulation and encoding schemes for passive systems should be selected carefully to reduce complexity while maintaining acceptable performance and stable power transfer to the transponder chip. ASK modulation is usually preferred for reader-to-transponder communications as it reduces the complexity of the transponder – a simple envelope detector circuit can be used to demodulate the incoming ASK-modulated signal in the transponder. To provide the transponder with enough power during reader-to-transponder data transmissions, the encoding scheme should avoid long periods of low RF signal. The pulse-interval encoding (PIE) scheme specified by ISO/IEC 18000-63 achieves this by using long RF carrier transmissions followed by short periods of low RF carrier transmissions. Another way to improve the energy transfer during reader-to-transponder data transmissions is by optimizing the reader modulation

index. Some protocols like ISO 14443-B allow the user to select the modulation index to best suit specific applications.

5. Backscatter modulator – uses a transistor switch to modulate the radar cross section of the transponder antenna by modulating the impedance presented to the antenna by the chip.

6. Memory – used to store transponder information, including the unique identification number (ID) and user data. Electrically erasable programmable read only memories (EEPROMs) are typically used and some devices allow parts of the memory (e.g., the transponder ID) to be programmed on the fly by the reader.

7. Digital control logic – performs vital tasks including baseband data encoding/decoding, communication protocol handling, and memory access. In low-cost transponders, digital control logic is implemented using simple finite state-machines, while advanced systems may use general-purpose microcontrollers (MCUs) [36], or more sophisticated resources such as field programmable gate arrays (FPGAs) and complex programmable logic devices (CPLDs) [37]. In RFID-enabled or passive wireless sensors, this section is also responsible for interfacing with internal and external sensors. Advanced devices may also include ADCs and digital input/output interfaces. For example, some devices incorporate an SPI interface to interact with an external MCU [33].

2.4.6   Chipless Transponders

Despite the numerous benefits of RFID such as long communication distance, no line-of-sight communication, ability to inventory multiple transponders automatically, and large transponder memory capacities, the deployment of the technology has been hindered by cost (mainly the unit cost of transponders).

Since the cost of RFID transponders is dominated by the cost of the integrated circuits, getting rid of the integrated circuits altogether has been proposed as a path toward lowering the cost of RFID. Chipless transponders or transponders that do not require integrated circuits has become an important emerging area of RFID technology [38–40]. Chipless transponders are usually built with low-cost discrete components or magnetic and electromagnetic reflective or absorptive materials. These transponders are simple and inexpensive and can be printed directly on products and packaging, benefiting from ubiquitous inkjet printing techniques in a similar way to barcodes. This technology has, however, two major drawbacks relating to the low data storage capacity of transponders and increased complexity of readers.

To encode and transfer data, chipless RFID systems use either time-domain reflectometry or frequency signature. In time-domain reflectometry, the reader broadcasts a pulse and listens for echoes from the transponder. In this method, the transponder data (usually a serial number) is encoded in the timing of pulse arrivals at the reader. In frequency signature, the reader sends a signal that spans a broad frequency range (a broadband pulse or a chirp signal) and monitors echoes in the frequency domain. Typically, the transponder data is encoded in the presence or absence of resonance peaks at predetermined frequencies of the echoed signal.

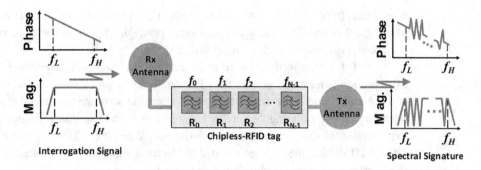

Figure 2.12 Chipless RFID system using frequency signature. In this scheme, the interrogation and echoed signals are separated by using different antenna polarizations. Reprinted from [38].

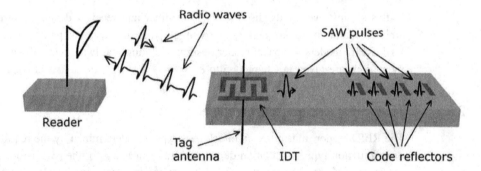

Figure 2.13 SAW-based RFID transponder. Reprinted from [41].

To attenuate or absorb energy at specific frequencies, transponders may use magnetic materials, chemicals, or resonant circuits.

An example of a 1-bit inductive chipless system using frequency signature has already been discussed in Section 2.4.1 (see Figure 2.7). A multi-bit system based on that scheme could be implemented by using multiple LC resonators tuned to different frequencies. The architecture of a chipless RFID transponder using frequency signature and electromagnetic coupling is presented in Figure 2.12, where several resonators are used to encode information bits. To detect the transponder resonances, a broadband pulse or a chirp signal is typically used.

2.4.7 Surface Acoustic Wave Transponders

A surface acoustic wave (SAW) transponder is a type of chipless transponder that uses time-domain reflectometry [41]. The key component of a SAW transponder is an interdigital transducer (IDT) consisting of two interlocking arrays of metallic electrodes deposited on the surface of a piezoelectric substrate (see Figure 2.13). By applying an AC electrical signal across the IDT electrodes, periodically distributed mechanical forces due to the piezoelectric effect create acoustic waves that propagate along the surface of the substrate. The same principle applies to the conversion of surface acoustic waves back to the electrical domain.

An IDT device, directly attached to the transponder antenna, converts an electrical interrogation pulse received from the RFID reader into a propagating surface acoustic wave. Narrow aluminum strips are disposed along the substrate to reflect the acoustic wave. By properly placing these reflectors, one can create a specific reflection pattern that encodes the desired transponder information. When the reflected acoustic waves hit the IDT device, they are reconverted to electrical signals that are retransmitted by the transponder antenna and detected and decoded by the reader.

2.5   UHF RFID Readers

In this section, we briefly discuss key transmitter and receiver design requirements, architectures, approaches, and tradeoffs involved in the design of passive-backscatter UHF RFID readers. Practical reader designs including hardware and software implementations are given in Chapters 5 to 8.

2.5.1   Transmitter Design Considerations

The RFID reader transmitter architecture is typically determined by the requirements of modulation type, modulation depth, and transmit mask for the reader-to-transponder transmissions. To comply with different modern protocols, UHF RFID reader transmitters should support single-sideband ASK (SSB-ASK), double-sideband ASK (DSB-ASK), and phase-reversal ASK (PR-ASK). To limit the transmit spectrum occupancy and comply with regulated spectrum masks, baseband pulse-shaping filters are typically required. In addition, the transmitter linearity should be optimized to minimize signal distortion and spectrum regrowth. Other key requirements of the transmitter design include linearity and efficiency, which are typically dominated by the transmit power amplifier stage. Another key aspect of the RFID transmitter design is the phase noise of the local oscillator; transmitter phase noise can leak into the receiver and be down-converted to baseband, degrading the reader sensitivity to the often-small transponder-backscattered signal, which usually is spectrally close to the reader-transmitted carrier (a few tens to a few hundred kilohertz).

2.5.2   Receiver Design Considerations

The RFID reader receiver architecture is determined by the data rate and the type of encoding and modulation of the transponder. Since passive UHF RFID transponders backscatter the carrier radiated by the reader to communicate their information, it is convenient to use the same local oscillator used in the transmitter to homodyne the signal backscattered from the transponders. In the next chapter we will have more to say about this type of receiver architecture.

Due to multipath propagation and the arbitrary distance between the reader and transponder, the reader local oscillator and the intercepted transponder signal generally present an arbitrary phase difference. Therefore, non-quadrature detection can yield null output when the local oscillator and received signal are out of phase. To prevent blind spots, readers commonly use quadrature direct conversion detection where the incoming transponder signal is cross-correlated with an in-phase and a quadrature component of the local oscillator signal. Hence, if the transponder signal is fully uncorrelated with one component of the local oscillator (i.e., phase-shifted by 90 degrees), it will be maximally correlated with the other component. This makes the detection of the transponder information insensitive to the relative phase of the transponder backscatter signal.

The direct conversion receiver architecture is prone to amplitude and phase imbalances and DC offsets due to transmitter-to-receiver leakage; in some cases DC offsets can be much bigger than the transponder baseband signal. This aggravates the dynamic range requirements of baseband circuitry. Fortunately, UHF RFID standards typically define encoding schemes with minimal DC components (e.g., FM0 and Miller coding, see Chapter 4), allowing the use of baseband filtering to eliminate unwanted DC offsets. DC offsets can also be mitigated by applying self-jamming suppression in the RF domain (see Chapter 8 for a comprehensive discussion on self-jamming). Other important design requirements include receiver sensitivity, noise figure, and dynamic range.

2.5.3   Typical UHF RFID Reader Architecture

Figure 2.14 depicts a simplified diagram of a typical UHF RFID reader based on a commercial application-specific integrated circuit (ASIC). The transmitter uses quadrature modulation to support various modulation schemes and the receiver employs quadrature demodulation to prevent blind reading spots that can occur due to the phase

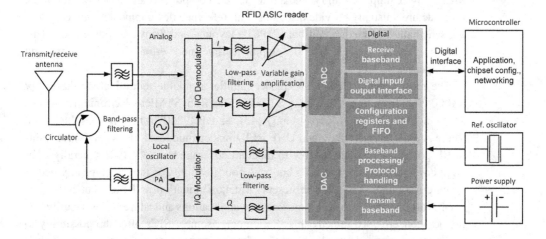

Figure 2.14 Simplified diagram of a typical ASIC-based UHF RFID reader.

uncorrelation mechanism described above. The complex baseband signal generated by the transmit digital signal processor is converted to the analog domain via two digital-to-analog converters (DACs) and then translated to RF via quadrature direct conversion modulation. The transponder RF signal intercepted by the reader is converted to baseband via quadrature direct conversion demodulation and the resulting complex baseband signal is then digitized by two ADCs and further processed to retrieve the transponder message.

In Figure 2.14, a circulator is used to share the antenna between transmitter and receiver, but a directional coupler could also be used for that purpose. To limit out-of-band emissions, the transmit power amplifier is usually followed by a bandpass filter centered at the RF carrier frequency. A bandpass filter can also be used to improve frequency selectivity in the receiver. UHF RFID readers may also employ low-noise amplifiers (LNAs) (omitted in Figure 2.14) for enhanced sensitivity, but this approach requires careful design as the LNA can be driven into saturation by self-jamming (more about self-jamming in Chapter 8).

In addition to the ASIC approach illustrated in Figure 2.14 [42–46], UHF RFID readers can be based on original equipment manufacturer (OEM) modules [47–49], discrete components [50–53], custom integrated circuits (ICs) [54–57], and software-defined radio (SDR) [58–65]. The choice of design approach depends on the desired flexibility, cost, and time to market. Chapters 5 through 8 discuss key aspects of UHF RFID reader engineering.

2.5.4   ASIC-Based Reader Design

Off-the-shelf ASIC readers integrate the main building blocks for radio-frequency synthesis, analog baseband processing, analog-to-digital and digital-to-analog conversion, digital baseband processing, and microcontroller interfacing. Therefore, ASICs require minimal external parts to assemble a UHF RFID reader – typically, a low-cost MCU, power supply circuitry, and a few discrete components are required. Moreover, manufacturers usually provide designers with reference development material including schematics, printed circuit board (PCB) layouts, and firmware source code. This can dramatically reduce the development time and effort for integrators and developers.

The most popular UHF ASIC reader families available today include the R500/R1000/2000 from Impinj Inc. [42], ST25RU399x from STMicroelectronics (formerly, AS399x from Austria Microsystems) [43], and PR9200 from Phychips Inc. [44]. These ASICs integrate the main UHF RFID reader functional blocks and have built-in RF power capabilities for low to medium coverage range. ASICs generally offer good cost–performance tradeoffs but have poor flexibility as they are typically tied to specific standards. The latter represents a major limitation as RFID standards have been evolving rapidly to accommodate new applications and address new security and privacy concerns. To remedy this shortcoming, some ASICs offer the possibility to bypass their protocol-specific baseband circuitry and allow the combination of their analog RF front end with external DSPs for custom baseband processing [43].

Table 2.5 UHF RFID reader design approaches and tradeoffs.

Design approach	Off-the-shelf ASIC	Custom IC	SDR
Development effort	Reduced	Involved	Involved
Required design skills	Reduced	Specialized	Specialized
Time to market	Short	—	Medium to long
Cost	Low to medium	—	Medium to high
Overall performance	Medium	Medium	Highest
Flexibility/upgradability	Limited	Reduced	Best

2.5.5 SDR-Based Reader Design

In SDR RFID reader implementations, some or all physical layer tasks (e.g., channel selection, tuning, modulation, and encoding), which are typically implemented in hardware, are performed in software. In this approach, the digital functions of the reader in Figure 2.14 would be implemented on a DSP or FPGA. The SDR design approach typically offers higher flexibility and performance compared to the ASIC approach.

The major benefit of SDR-based reader systems is their ability to accommodate new standards through software upgrade, without the need for hardware redesign. Flexibility is also what makes SDR implementations ideal for RFID research, characterization, measurement, and protocol evaluation [58, 59, 63, 64]. But the improved SDR flexibility comes at the expense of increased complexity, cost, and time to market compared to commercial ASIC-based designs.

Commercially available SDR-based reader solutions include the Speedway reader from Impinj Inc. [66] and the Enterprise ALR-F800 reader from Alien Technology [67]. Table 2.5 summarizes the main UHF RFID reader design approaches and the tradeoffs involved. In Chapter 7, we report on the full-stack design of a cost-effective SDR RFID reader.

References

[1] R. Das, "RFID Forecasts, Players and Opportunities 2019–2029: The Complete Analysis of the Global RFID Industry," IDTechEx research report, 2019.
[2] D. M. Dobkin, *The RF in RFID: Passive UHF in Practice*, 2nd ed. Newnes, Oxford, 2012.
[3] K. Finkenzeller, *RFID Handbook: Fundamentals and Applications in Contactless Smart Cards, Radio Frequency Identification and Near-Field Communication*, 3rd ed. Wiley, Chichester, 2010.
[4] P. Pursula, et al., "Wirelessly Powered Sensor Transponder for UHF RFID," in *Proceedings of Transducers '07 & Eurosensors XXI Conference*, 2007, pp. 73–76.
[5] J. R. Smith, *Wirelessly Powered Sensor Networks and Computational RFID*, Springer, New York, 2013.
[6] C. A. Balanis, *Antenna Theory: Analysis Design*, 3rd ed. John Wiley & Sons, Hoboken, NJ, 2005.

[7] K. Sattlegger and U. Denk, "Navigating Your Way Through the RFID Jungle." White paper SLYY056, Texas Instruments, Dallas, TX, 2014.

[8] M. Ilyas and S. A. Ahson, "*RFID Handbook: Applications, Technology, Security, and Privacy.*" CRC Press, Boca Raton, FL, 2008.

[9] IEEE-USA, "The State of RFID Implementation and Its Policy Implications: An IEEE-USA White Paper," 2009.

[10] RFIDWorld, "Levi Strauss & Co. Teams Up with Intel to Test RFID in Stores." [Online], available at: https://rfidworld.ca/levi-strauss-co-teams-up-with-intel-to-test-rfid-in-stores/2429

[11] C.-J. Li, L. Liu, S.-Z. Chen, C. C. Wu, C.-H. Huang, and X.-M. Chen, "Mobile Healthcare Service System Using RFID," in *IEEE International Conference on Networking, Sensing and Control*, vol. 2, 2004, pp. 1014–1019.

[12] C. Seckman, A. Bauer, T. Moser, and S. Paaske, "The Benefits and Barriers to RFID Technology in Healthcare," *Online Journal of Nursing Informatics*, 21(2), 2017.

[13] D. Dixit, A. Kalbande, and K. M. Bhurchandi, "RFID-Based Health Assistance & Monitoring System Through a Handmounted Embedded Device," in *2013 Fourth International Conference on Computing, Communications and Networking Technologies (ICCCNT)*, 2013, pp. 1–9.

[14] A.P. Hridhya, C. Periasamy, and I.R Rahul, "Patient Monitoring and Abnormality Detection Along with an Android Application," *International Journal of Computer Communication and Informatics*, 1(1):52–57, 2019.

[15] Airbus Aerospace Company, "Airbus Selects Fujitsu RFID Integrated Label Technology for Management, Traceability of Aircraft Parts," May 2014. [Online], available at: www.intelligent-aerospace.com/commercial/article/16541615/airbus-selects-fujitsu-rfid-integrated-label-technology-for-management-traceability-of-aircraft-parts

[16] Boeing Aerospace Company, "RFID Integrated Solutions System Optimizes Maintenance Efficiency." [Online], available at: www.boeing.com/commercial/aeromagazine/articles/2012_q1/2/

[17] M. Roberti, "Airline Industry Embraces RFID Baggage Tracking," RFID Journal, June 2018. [Online], available at: www.rfidjournal.com/airline-industry-embraces-rfid-baggage-tracking

[18] International Air Transport Association, "Resolution: RFID Baggage Tracking Set for Global Deployment," June 2019. [Online], available at: www.iata.org/en/pressroom/pressroom-archive/2019-press-releases/2019-06-02-05/

[19] Wikipedia, "Biometric Passport." [Online], available at: https://en.wikipedia.org/wiki/Biometric_passport

[20] GS1, "Regulatory Status for Using RFID in the EPC Gen2 (860 to 960 MHz) Band of the UHF Spectrum," February 11, 2020. [Online], available at: www.gs1.org/docs/epc/uhf_regulations.pdf

[21] U.S. Federal Communications Commission, "Part 15.247." [Online], available at: www.govinfo.gov/content/pkg/CFR-2013-title47-vol1/pdf/CFR-2013-title47-vol1-sec15-247.pdf https://www.law.cornell.edu/cfr/text/47/15.247

[22] ETSI, "Short Range Devices (SRD) Operating in the Frequency Fange 25 MHz to 1000 MHz; Part 2: Harmonised Standard Covering the Essential Requirements of Article 3.2 of Directive 2014/53/EU for Non Specific Radio Equipment," EN 300 220-2, V3.1.1, 2017. [Online], available at: www.etsi.org/deliver/etsi_en/300200_300299/30022002/03.01.01_60/en_30022002v030101p.pdf

[23] RFID4U, "RFID Basics – RFID Regulations." [Online], available at: https://rfid4u.com/rfid-regulations/

[24] ETSI, "Radio Frequency Identification Equipment Operating in the Band 865 MHz to 868 MHz with Power Levels up to 2 W and in the Band 915 MHz to 921 MHz with Power Levels up to 4 W; Harmonised Standard for Access to Radio Spectrum," EN 302 208 C3.3.0, 2020. [Online], available at: www.etsi.org/deliver/etsi_en/302200_302299/302208/03.03.00_20/en_302208v030300a.pdf

[25] European Commission, "Commission Implementing Decision (EU) 2017/1483 of 8 August 2017." [Online], available at: https://eur-lex.europa.eu/legal-content/EN/TXT/PDF/?uri=CELEX:32017D1483&from=en

[26] European Commission, "Commission Implementing Decision (EU) 2018/1538 of 11 October 2018." [Online], available at: https://eur-lex.europa.eu/legal-content/EN/TXT/PDF/?uri=CELEX:32018D1538&from=EN

[27] M. Lenehan, "EU Upper Band Country List," Impinj. [Online], available at: https://support.impinj.com/hc/en-us/articles/360006823440-EU-Upper-Band-Country-List

[28] International Standards Organization, "ISO/IEC 14443-2:2010, Identification Cards – Contactless Integrated Circuit Cards – Proximity Cards – Part 2: Radio Frequency Power and Signal Interface." [Online], available at: www.iso.org/standard/50941.html

[29] Ecma International, "ECMA-352: Near Field Communication Interface and Protocol – 2 (NFCIP-2)," 4th ed., 2021. [Online], available at: https://ecma-international.org/publications-and-standards/standards/ecma-352/

[30] International Standards Organization, "ISO/IEC 15693-2:2006, Identification Cards – Contactless Integrated Circuit Cards – Vicinity Cards – Part 2: Air Interface and Initialization." [Online], available at: www.iso.org/standard/39695.html

[31] H. Cravo Gomes and N. Borges Carvalho, "The Use of Intermodulation Distortion for the Design of Passive RFID," in Proceedings of the 37th European Microwave Conference, pp. 377–380, 2007.

[32] S. J. Thomas, et al., "Quadrature Amplitude Modulated Backscatter in Passive and Semipassive UHF RFID Systems," *IEEE Transactions on Microwave Theory and Techniques*, 60(4):1175–1182, 2012.

[33] AMS, "SL900A EPC Gen2 Sensor Tag." [Online], available at: www.ams.com/SL900A.

[34] Y. Wang, et al., "Design of a Passive UHF RFID Tag for the ISO18000–6C Protocol," *Journal of Semiconductors*, 32(5):055009, 2011.

[35] M. Baghaei-Nejad, et. al., "A Remote-Powered RFID Tag with 10Mb/s UWB Uplink and −18.5dBm Sensitivity UHF Downlink in 0.18μm CMOS," in *2009 IEEE International Solid-State Circuits Conference – Digest of Technical Papers*, 2009, pp. 198–199a.

[36] J. R. Smith, et al., "A Wirelessly-Powered Platform for Sensing and Computation," in *8th International Conference on Ubiquitous Computing (Ubicomp)*, 2006, pp. 495–506.

[37] S. Roy, et al., "RFID: From Supply Chains to Sensor Nets," *Proceedings of the IEEE*, 98 (9):1583–1592, 2010.

[38] C. Herrojo, et al., "Chipless-RFID: A Review and Recent Developments," *Sensors*, 19 (15):3385, 2019.

[39] S. Preradovic and N. C. Karmakar, "Chipless RFID: Bar Code of the Future," *IEEE Microwave Magazine*, 11(7):87–97, 2010.

[40] N. C. Karmakar, E. M. Amin, and J. K. Saha, *Chipless RFID Sensors*. John Wiley & Sons, Inc., Hoboken, NJ, 2016.

[41] V. P. Plessky and L. M. Reindl, "Review on SAW RFID Tags," *IEEE Transactions on Ultrasonics, Ferroelectrics, and Frequency Control*, 57(3):654–668, 2010.

[42] Impinj, "Indy R2000 UHF Gen 2 RFID Reader Chip." [Online], available at: www.impinj.com

[43] STMicroelectronics, "ST25RU3993, UHF RFID Single Chip Reader EPC Class1 Gen2 Compatible." [Online], available at: www.st.com

[44] PHYCHIPS, "PR9200, UHF RFID Reader SoC." [Online], available at: www.phychips.com

[45] M. Koller and R. Küng, "Adaptive Carrier Suppression for UHF RFID Using Digitally Tunable Capacitors," in *2013 European Microwave Conference*, 2013, pp. 943–946.

[46] O. Bostan, H. U. Aydoğmuş, and S. Topaloğlu, "Design of a Low-Power, Low-Cost UHF RFID Reader Module," *Turkish Journal of Electrical Engineering and Computer Sciences*, 24(4):2747–2755, 2016.

[47] JADAK, "M6E-NANO, Mercury6e UHF RFID Module Family Performance, Efficiency, and Flexibility, ThingMagic Technology." [Online], available at: www.jadaktech.com

[48] Sanray, "M2210 UHF RFID module." [Online], available at: www.sanrayrfid.com

[49] Impinj, "RS500/RS1000/RS2000 RAIN RFID Reader Module Family." [Online], available at: www.impinj.com

[50] P. Pursula, M. Kiviranti, and H. Seppa, "UHF RFID Reader with Reflected Power Canceller," *IEEE Microwave and Wireless Components Letters*, 19(1):48–50, 2009.

[51] B. You, B. Yang, X. Wen, and L. Qu, "Implementation of Low-Cost UHF RFID Reader Front-Ends with Carrier Leakage Suppression Circuit," *International Journal of Antennas and Propagation*, 2013:135203, 2013.

[52] P. Nikitin, S. Ramamurthy, and R. Martinez, "Simple Low Cost UHF RFID Reader," in *Proc. 2013 IEEE International Conference on RFID*, 2013.

[53] A. Povalac, "Experimental Front End for UHF RFID Reader," *Elektro Revue*, 2(1):55–59.

[54] S. Chiu et al., "A 900 MHz UHF RFID Reader Transceiver IC," *IEEE Journal of Solid-State Circuits*, 42(12):2822–2833, 2007.

[55] R. Zhang et al., "A Single-Chip CMOS UHF RFID Reader Transceiver," in *2010 IEEE Radio Frequency Integrated Circuits Symposium*, 2010, pp. 101–104.

[56] L. Ye et al., "A Single-Chip CMOS UHF RFID Reader Transceiver for Mobile Applications," in *2009 Proceedings of ESSCIRC*, 2009, pp. 228–231.

[57] N. Usachev et al., "System Design Considerations of Universal UHF RFID Reader Transceiver ICs," *Facta Universitatis, Electronics and Energetics*, 28(2):297–307, 2015.

[58] N. Kargas, F. Mavromatis, and A. Bletsas, "Fully-Coherent Reader with Commodity SDR for Gen2 FM0 and Computational RFID," *IEEE Wireless Communications Letters*, 4(6):617–620, 2015.

[59] L. Catarinucci et al., "A Cost-Effective SDR Platform for Performance Characterization of RFID Tags," *IEEE Transactions on Instrumentation and Measurement*, 1(4):903–911, 2012.

[60] W. Yuechun et al., "A Flexible Software Defined Radio-Based UHF RFID Reader Based on the USRP and LabView," in *2016 International SoC Design Conference (ISOCC)*, 2016, pp. 217–218.

[61] V. Yang, E. Zhang, and A. He, "Developing a UHF RFID Reader RF Front End with an Analog Devices' Solution," Analog Devices, Inc. technical article, 2019.

[62] A. Oliveira et al., "All-Digital RFID Readers: An RFID Reader Implemented on an FPGA Chip and/or Embedded Processor," *IEEE Microwave Magazine*, 22(3):18–24, 2021.

[63] M. Buettner and D. Wetherall, "A Software Radio-Based UHF RFID Reader for PHY/MAC Experimentation," in *IEEE International RFID Conference*, 2011, pp. 134–141.

[64] A. Povalac, "Spatial Identification Methods and Systems for RFID Tags," Ph.D. thesis, Brno University of Technology, 2012.

[65] E. A. Keehr, "A Low-Cost Software-Defined UHF RFID Reader with Active Transmit Leakage Cancellation," in *2018 IEEE International Conference on RFID*, 2018, pp. 1–8.

[66] Impinj, "Speedway Reader." [Online], available at: www.impinj.com

[67] Alien Technology, "Enterprise ALR-F800 UHF RFID Reader." [Online], available at: www.alientechnology.com

3   Backscatter Communications: Fundamentals and Recent Advances

Scattering of electromagnetic waves from objects was first explored during World War II for aircraft detection, and deliberate reflection of radio waves as a means of communicating information was proposed for the first time by H. Stockman in 1948 [1]. In radar theory, an object can be characterized by its radar cross section, a measure of how reflective the object is to electromagnetic waves. In modern RFID, cooperative radar or deliberate wave reflection is combined with wireless power transmission to remotely retrieve data from passive transponders.

Backscatter transponders have no local RF oscillators and instead modulate their antenna radar cross section to reflect radio signals from an RFID interrogator with a specific signature that encodes the message to be transmitted to the reader (see Figure 3.1). Besides not requiring a local radio oscillator, passive-backscatter devices have no onboard power supply and harvest DC power to supply their electronics from the signal radiated by the reader. One of the first attempts to combine backscatter and wireless power transmission dates to 1960 and is credited to D. B. Harris who proposed a voice transmission system that required no batteries in the mobile device [2].

In this chapter we (i) discuss underlying concepts of backscatter communications, including transponder effective antenna aperture and radar cross section modulation; (ii) perform a power budget analysis for a typical passive-backscatter system, derive the backscatter power, and determine major backscatter range limiting factors; (iii) describe backscatter demodulation methods used in passive UHF RFID systems; (iv) discuss recent advances in backscatter communications such as quadrature amplitude (QAM) modulation and multicarrier backscatter modulation; and (v) review state-of-the-art techniques used to enhance wireless power transfer, communication range, and the overall performance of passive-backscatter systems.

3.1   Antenna Thévenin-Equivalent Model

A typical passive backscatter transponder consists of an RFID integrated circuit / chip attached to a half-wavelength dipole antenna (Figure 3.2(a)). The transponder circuit can be modeled by the Thévenin-equivalent circuit of the antenna in series with the load impedance of the chip (Figure 3.2(a)). The antenna impedance (3.1) comprises a radiation resistance R_{rad}, which models the radio power radiated by the antenna, a loss

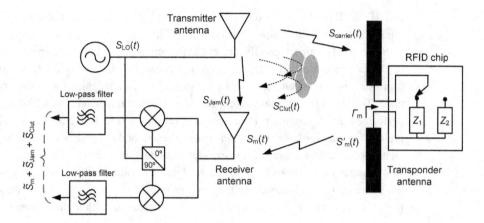

Figure 3.1 Illustration of backscatter communication. Left: Interrogator transceiver front end. Right: Transponder backscatter front end.

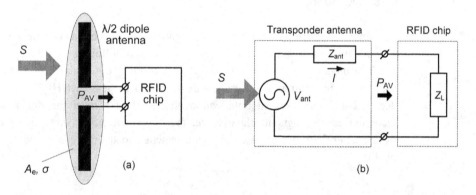

Figure 3.2 (a) Passive backscatter transponder comprising RFID chip and dipole antenna. (b) Corresponding model consisting of Thévenin equivalent of the antenna cascaded with equivalent impedance of the chip.

resistance R_{ant}, which accounts for losses in the antenna, and a reactance X_{ant}, which describes the reactive component of the antenna [3]. The RFID chip can be modeled by its complex input impedance (3.2):

$$Z_{ant} = R_{rad} + R_{ant} + jX_{ant}, \qquad (3.1)$$

$$Z_L = R_L + jX_L, \qquad (3.2)$$

where j denotes the imaginary unit. At resonance, the antenna presents minimal (ideally zero) reactance, and for low-loss antennas the impedance can be approximated by the radiation resistance R_{rad}. To maximize the power transferred to the RFID chip, the antenna impedance should match the complex conjugate impedance of the chip. In passive transponders, conjugate impedance matching is performed for the high-impedance (or absorptive) state of the chip. A survey of the design and matching of UHF RFID antennas is presented in [4], and an approach to antenna design and integration with RFID chips is described in [5].

Since RFID chips typically present a power- and frequency-dependent response [6], impedance-matching design should account for the operating conditions that transponders will encounter in real scenarios as these can significantly affect performance. To maximize communication range, conjugate impedance matching is usually performed at the minimum input power level required to activate the chip [7]. This typically occurs for large separation between reader and transponder, or due to line-of-sight blockage and degraded transponder antenna gain caused by object attachment. For stronger excitations closer to the reader, transponders will generally work even under mismatch conditions. But big variations in antenna impedance under high power levels can render transponders powerless even within the operational read range [6].

3.2   Effective Aperture

To study the effective aperture and radar cross section of the transponder antenna, we use the model of Figure 3.2(b) and follow a derivation similar to [3]. The model in Figure 3.2(b) has some important limitations [8–11] but can be applied to the minimum-scattering antennas used in passive UHF RFID transponders [12–14]. For optimal polarization alignment of a receiving antenna with an incoming electromagnetic plane wave, the maximum power that can be delivered by the antenna to a complex conjugate matched load or available power is given by [3]

$$P_{av} = SA_e, \tag{3.3}$$

where S is the power flux density describing the amount of power per unit of area carried by the incoming electromagnetic field, and A_e is the effective aperture of the antenna representing the functional area through which the available power P_{av} passes when the antenna is exposed to the power flux density S perpendicular to the direction of propagation.

The incoming electromagnetic field induces a root-mean-square voltage across the terminals of the transponder antenna, V_{ant}, and a root-mean-square current through the circuit, I, such that

$$I = \frac{V_{ant}}{|Z_{ant} + Z_L|} = \frac{V_{ant}}{\sqrt{(R_{rad} + R_{ant} + R_L)^2 + (X_{ant} + X_L)^2}}. \tag{3.4}$$

Substituting (3.4) into the definition of the power delivered to the load $\left(P_{av} = I^2 R_L\right)$ leads to

$$P_{av} = \frac{V_{ant}^2 R_L}{(R_{rad} + R_{ant} + R_L)^2 + (X_{ant} + X_L)^2}. \tag{3.5}$$

Combining (3.3) and (3.5) yields the antenna effective aperture,

$$A_e = \frac{P_{av}}{S} = \frac{V_{ant}^2 R_L}{\left[(R_{rad} + R_{ant} + R_L)^2 + (X_{ant} + X_L)^2\right]S}.$$
(3.6)

Assuming a lossless transponder antenna ($R_{ant} = 0$) and conjugate impedance matching ($R_{rad} = R_L$ and $X_{ant} = -X_L$), (3.6) simplifies to

$$A_e = A_{emax} = \frac{V_{ant}^2}{4SR_{rad}}.$$
(3.7)

Equation (3.7) represents the total area through which power is collected from the incoming electromagnetic field and delivered to the load. The previous derivation assumes the common configuration found in off-the-shelf transponders where the antenna is directly connected to the load, but if the antenna and load are interconnected via a transmission line, then Z_L must be replaced with the input impedance looking into the transmission line at the antenna location.

3.3 Scattering Antenna Aperture

Another important characteristic of an antenna is the radar cross section, σ, which determines the power scattered from the antenna. The radar cross section can be defined in a similar way to above by realizing that: (i) the current I induced by the incoming electromagnetic field in the circuit in Figure 3.2(b) flows through the radiation resistance R_{rad}, and (ii) this current has the same effect as a current created by a transmit generator connected to the antenna. Thus, the power "dissipated" in the radiation resistance, due to the current I, corresponds to the re-radiated power P_{scat} given by

$$P_{scat} = I^2 R_{rad}.$$
(3.8)

As previously done for (3.6), we define the scattering aperture or radar cross section as

$$\sigma = \frac{P_{scat}}{S} = \frac{I^2 R_{rad}}{S} = \frac{V_{ant}^2 R_{rad}}{\left[(R_{rad} + R_{ant} + R_L)^2 + (X_{ant} + X_L)^2\right]S}.$$
(3.9)

Assuming again a lossless antenna and conjugate impedance matching, (3.9) simplifies to

$$\sigma = \frac{V_{ant}^2}{4SR_{rad}}.$$
(3.10)

This result indicates that, under conjugate matching, the antenna re-radiates as much power as it delivers to the load (i.e., $\sigma = A_{emax}$). Figure 3.3 presents the effective aperture (3.6) and radar cross section (3.9) relative to the maximum effective aperture (3.7) as a function of normalized load resistance ($r_L = R_L/R_{rad}$) [3, 15].

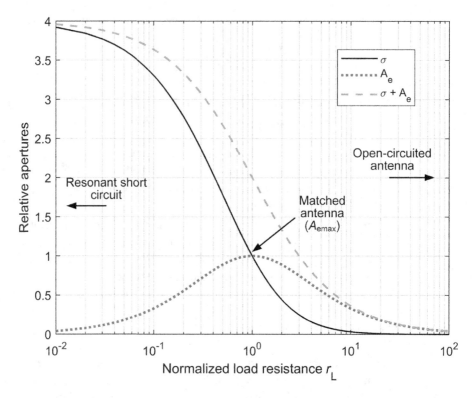

Figure 3.3 Relative effective aperture and radar cross section as a function of normalized load resistance for $R_{ant} = X_{ant} = X_L = 0$. The intersection between the aperture and radar cross section curves corresponds to the conjugate matched condition.

From the previous analysis, we can identify the following special regimes.

1. Conjugate-matched antenna ($R_L = R_{rad}$ and $X_L = -X_{ant}$): The radar cross section equals the maximum effective aperture ($\sigma = A_{emax}$), and the antenna scatters the same amount of power that is delivered to the load.
2. Resonant short-circuited antenna ($R_L = 0$ and $X_L = -X_{ant}$): Four times as much power is re-radiated as under the previous conjugate matched condition ($\sigma = 4A_{emax}$).
3. Open-circuited antenna ($Z_L = \infty$): No current flows through the circuit and the system presents null effective aperture and radar cross section ($\sigma = A_e = 0$). Ideally no power gets re-radiated, but in practice residual re-radiation occurs due to parasitic components [12–14].

The results above suggest that the resonant short-circuit regime should be preferred to maximize backscatter, but this regime does not provide maximum re-radiated power. In fact, if the antenna impedance is sufficiently reactive, the conjugate-matched antenna can scatter more power than the short-circuited antenna (see Figure 3.4) [14]. Note that the analysis presented here only considers relative power. Expressions for the absolute power scattered from and collected by the transponder antenna are given in Section 3.6.

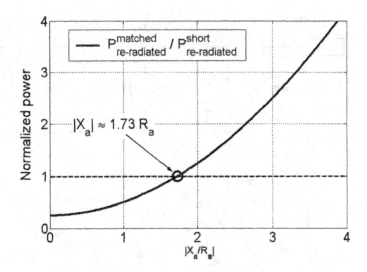

Figure 3.4 Power re-radiated by conjugate-matched antenna relative to short-circuited antenna as a function of normalized antenna reactance. Reprinted from [14].

3.3.1 Modulating the Scattered Electromagnetic Field

The loaded antenna properties discussed above are used in backscatter RFID systems to convey information from the transponder to the reader. By placing a switch device (typically a transistor) across the antenna in Figure 3.2(b), one can create a backscatter front end with two modulation states that can be used to communicate binary information.

In modulation state 1, with the switch open as in Figure 3.5(a), the antenna is conjugate matched to the chip, allowing maximum power absorption while some background power is scattered by the antenna. In modulation state 2 we close the switch, short-circuiting the antenna as illustrated in Figure 3.5(b). The power scattered from the antenna in this state is increased compared to the background state (see Figure 3.5(c)). The following derivations use an approach similar to [16].

During modulation state 1, a current $I_1 = V_{ant}/(2R_{rad})$ flows through the circuit (Figure 3.5(a)) and power P_{scat} is scattered by the antenna for a purely real chip impedance ($R_L = R_{ant}$),

$$P_{scat1} = P_{L1} = \frac{V_{ant}^2}{4R_{rad}}. \tag{3.11}$$

Equation (3.11) gives the maximum power delivered to the chip, P_{av}, which equals the transponder background scattered power. During modulation state 2, a current $I_2 = V_{ant}/R_{rad}$ flows through the circuit (Figure 3.5(b)), and four times as much power as the background power is scattered,

$$P_{scat2} = \frac{V_{ant}^2}{R_{rad}} = 4P_{av}. \tag{3.12}$$

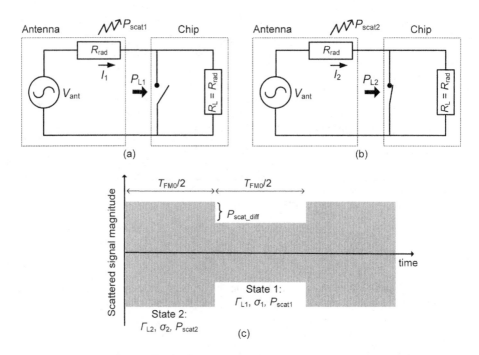

Figure 3.5 (a) Modulation state 1 with switch open and transponder antenna conjugate matched to the RFID chip. (b) Modulation state 2 with switch closed and no power delivered to the chip. (c) Illustration of waveform for FM0 transponder encoding. Adapted from [16].

During modulation state 2, no power is delivered to the chip because the voltage drop across the chip is zero, $P_{L2} = 0$. For an encoding scheme that switches symmetrically between two modulation states with equal probabilities (e.g., FM0 or Miller subcarrier), the average power delivered to the chip is

$$\overline{P_L} = \frac{P_{L1} + P_{L2}}{2} = \frac{P_{av}}{2}. \tag{3.13}$$

Based on the difference current flowing in the circuit in the two modulation states, $I_{diff} = |I_1 - I_2| = V_{ant}/(2R_{rad})$, one can determine the differential power scattered by the transponder:

$$P_{diff} = I_{diff}^2 R_{rad} = \frac{V_{ant}^2}{4R_{rad}} = P_{av}. \tag{3.14}$$

Considering again symmetrical encoding and equally likely modulation states, the average scattered power is

$$\overline{P_{diff}} = \frac{P_{av}}{2}. \tag{3.15}$$

A similar analysis can be done for different load impedance combinations (e.g., matched/open or short/open) [16]. Results for different modulation schemes are

Table 3.1 Power budget for elementary modulation schemes.

Modulation scheme	Z_{L1}	Z_{L2}	Γ_{L1}	Γ_{L2}	P_{scat} (mean)	P_L (mean)
Scheme 1 (matched/short)	Z_{ant}^*	0	0	-1	$P_{av}/2$	$P_{av}/2$
Scheme 2 (open/matched)	∞	Z_{ant}^*	1	0	$P_{av}/2$	$P_{av}/2$
Scheme 3 (short/open)	0	∞	-1	1	$2P_{av}$	0

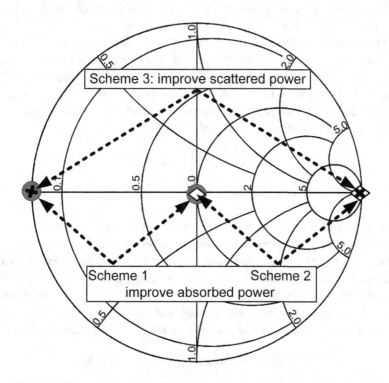

Figure 3.6 Illustration of the tradeoff between transponder scattered and absorbed power.

summarized in Table 3.1 and the modulation scheme choice tradeoff is illustrated in Figure 3.6. While modulation scheme 3 in Figure 3.6 maximizes the scattered power, it is not suitable for a fully passive transponder as it provides no power to the chip. But this scheme could well be used in a battery-assisted or semi-passive transponder with an ancillary power supply. The reflection coefficient presented in Table 3.1 for an antenna directly attached to a load is given by the Kurokawa reflection coefficient (3.16). For antennas with purely real impedance, the Kurokawa reflection coefficient reduces to the standard reflection coefficient definition [3],

$$\Gamma_L = \frac{Z_L - Z_{ant}^*}{Z_L + Z_{ant}}. \tag{3.16}$$

3.4   Quadrature Backscatter Demodulation

For the description of the demodulation method typically used in backscatter systems, observe the basic system of Figure 3.1, where we neglect the effect of noise, assume unitary amplitude scaling factors in the communication channel, and refer to electric fields and circuit voltages interchangeably as signals, $S(t)$. In addition, all signals are phase-referenced to the phase of the reader local oscillator, and the communication channel is assumed to be stationary, that is, it remains unchanged during each reader–transponder transaction.

The reader broadcasts a carrier signal that serves both as remote power supply and oscillator for the transponder. By modulating its antenna radar cross section, the transponder imprints a signature in the reflected signal to communicate its information. Passive UHF RFID readers typically employ quadrature direct conversion to demodulate the transponder backscattered signal. The carrier broadcast by the reader and received by the transponder can be written as

$$S_{\text{carrier}}(t) = \text{Re}\left\{A_{\text{carrier}}e^{(j\omega t + \phi_{\text{carrier}})}\right\}, \tag{3.17}$$

where $\text{Re}\{\cdot\}$ represents the real part of a complex variable, ω is the operating frequency, and A_{carrier} and Φ_{carrier} are the amplitude and phase of the reader carrier at the transponder location, respectively. The transponder modulates its antenna radar cross section by loading the antenna with an impedance Z_m that is varied over time to encode information in the re-radiated carrier. The scattering from a transponder antenna has some subtleties [3, 4], but for the purpose of explaining the basic backscatter demodulation method, it can be characterized by a complex reflection coefficient Γ_m. The transponder backscatters the reader carrier (3.17) yielding the backscattered signals at the transponder location (3.18) and reader location (3.19):

$$S'_m(t) = \text{Re}\left\{\rho_m A_{\text{carrier}}e^{(j\omega t + \phi_{\text{carrier}} + \phi_m)}\right\}, \tag{3.18}$$

$$S_m(t) = \text{Re}\left\{\rho_m A_{\text{carrier}}e^{(j\omega t + 2\phi_{\text{carrier}} + \phi_m)}\right\}, \tag{3.19}$$

where ρ_m and Φ_m are the amplitude and phase of the transponder-modulated reflection, respectively, and Φ_{carrier} is a one-way modulation-independent phase introduced by the radio channel. The reader transmitter-to-receiver leakage due to self-jamming in the reader and environment clutter reflections can be expressed as

$$S_{\text{Jam}}(t) = \text{Re}\left\{A_{\text{Jam}}e^{(j\omega t + \phi_{\text{Jam}})}\right\}, \tag{3.20}$$

$$S_{\text{Clut}}(t) = \text{Re}\left\{\sum_{n=1}^{N} \rho_n A_{\text{carrier}}e^{(j\omega t + \phi_n)}\right\}, \tag{3.21}$$

where A_{Jam} and Φ_{Jam} are the amplitude and phase of the reader self-jamming signal, respectively, and ρ_n and Φ_n are the amplitudes and phases of the clutter reflection components, respectively. For simplicity, and assuming linearity, we process the signals (3.19) to (3.21) separately in the quadrature demodulator of Figure 3.1 and then combine the resulting baseband signals. By correlating the incoming signal (3.19) with the reader in-phase and quadrature local oscillator over one baseband symbol T_{symbol}, we obtain the complex baseband envelope corresponding to the modulation state m:

$$
\begin{aligned}
\widetilde{S_m} &= S_m^I + jS_m^Q \\
&= \frac{1}{T_{\text{symbol}}} \int_{-T_{\text{symbol}}/2}^{T_{\text{symbol}}/2} \operatorname{Re}\left\{\rho_m A_{\text{carrier}} e^{(j\omega t + 2\phi_{\text{carrier}} + \phi_m)}\right\} A_{\text{carrier}} \cos(\omega t) dt \\
&\quad + j \frac{1}{T_{\text{symbol}}} \int_{-T_{\text{symbol}}/2}^{T_{\text{symbol}}/2} \operatorname{Re}\left\{\rho_m A_{\text{carrier}} e^{(j\omega t + 2\phi_{\text{carrier}} + \phi_m)}\right\} A_{\text{carrier}} \sin(\omega t) dt \\
&= \frac{\rho_m A_{\text{carrier}}^2}{2} \cos(\phi_m + 2\phi_{\text{carrier}}) + j \frac{\rho_m A_{\text{carrier}}^2}{2} \sin(\phi_m + 2\phi_{\text{carrier}}).
\end{aligned}
\tag{3.22}
$$

This operation corresponds to a Fourier transform with an integration window of one transponder symbol period with fixed baseband modulation state. The resulting complex baseband envelope of the reader-intercepted signal can be written using more compact notation:

$$
\widetilde{S_m} = \frac{\rho_m A_{\text{carrier}}^2}{2} e^{j(2\phi_{\text{carrier}} + \phi_m)}.
\tag{3.23}
$$

Using a similar approach, one can obtain the complex baseband components due to self-jamming and clutter:

$$
\widetilde{S_{\text{Jam}}} = \frac{A_{\text{Jam}} A_{\text{carrier}}}{2} e^{j\phi_{\text{Jam}}},
\tag{3.24}
$$

$$
\widetilde{S_{\text{Clut}}} = \frac{A_{\text{carrier}}^2}{2} \sum_{n=1}^{N} \rho_n e^{j\phi_n}.
\tag{3.25}
$$

The combined baseband signal for the modulation state m becomes

$$
\begin{aligned}
\widetilde{S_m^T} &= \widetilde{S_m} + \widetilde{S_{\text{Jam}}} + \widetilde{S_{\text{Clut}}} \\
&= \frac{\rho_m A_{\text{carrier}}^2}{2} e^{j(\phi_m + 2\phi_{\text{carrier}})} + \frac{A_{\text{Jam}} A_{\text{carrier}}}{2} e^{j(\phi_{\text{Jam}})} + \frac{A_{\text{carrier}}^2}{2} \sum_{n=1}^{N} \rho_n e^{j(\phi_n)}.
\end{aligned}
\tag{3.26}
$$

3.4.1 Impact of Self-Jamming and Clutter

The first term in (3.26) contains the transponder baseband information, and the last two terms are complex DC offset components that depend on self-jamming and clutter signals. The model described above is simplified to facilitate understanding of the basic backscatter communication mechanism. In practice, there are important non-idealities that make backscatter communication challenging. For example, typical RFID channels are not stationary and as a result direct conversion receivers can produce time-varying DC offsets. In Chapter 8, we discuss techniques to deal with self-jamming and DC offsets, including adaptive suppression algorithms that can track changes in the RFID channel. This model also assumes a noiseless communication channel. In real applications, white noise in the channel and phase noise from the transmitter can find their way into the receiver. The latter does so through self-jamming and clutter reflections. Clutter phase noise is especially problematic in that it cannot be cancelled satisfactorily by a first-order cancelling network due to potentially large delays between the clutter scatterer and the reader receiver.

3.4.2   BPSK Backscatter Demodulation

In binary phase-shift keying (BPSK) backscatter modulation, the reflection coefficient amplitude remains constant $(\rho_1 = \rho_2)$ while the phase is flipped by 180° between the two modulation states $\left(\Phi_2 = \Phi_1 + 180^\circ\right)$. Figures 3.7(b)–(d) illustrate instances of (3.26) for BPSK modulation.

A trivial example of a BPSK constellation produced at a transponder antenna is shown in Figure 3.7(a) and the corresponding baseband constellation received by the reader is illustrated in Figures 3.7(b)–(d). The backscattered signal undergoes arbitrary phase rotations induced by the radio channel, and the down-converted baseband signal generally has a DC offset due to self-jamming. By eliminating the complex DC offset in the received baseband signal, the received constellation is centered about the origin of the complex plane (Figure 3.7(c)).

The resulting constellation after DC removal corresponds to a BPSK modulation (Figure 3.7(c)) and can be demodulated as such (demodulation algorithms are detailed in Chapters 5–7). BPSK symbol mapping is illustrated in Figure 3.7(d), where a binary decision is made by comparing the BPSK symbol phase to a decision threshold. Note that in practice transponders present impedance modulation states with resistive and capacitive characteristics, which leads to mixed ASK/PSK modulation (see Figure 3.8) [17]. But by centering the received constellation, the signal can be demodulated as a BPSK signal. In reality, noise and other non-idealities in the channel lead to signal constellations resembling clouds of points rather than just localized points as illustrated here. In Chapter 5, we provide scripts for evaluating these effects more realistically.

3.5   Advanced Backscatter Communication Techniques

This section discusses advances in backscatter communications geared toward increasing data rate and performance at reduced energy consumption and complexity. The techniques discussed here include quadrature backscatter modulation, spread-spectrum backscatter modulation, multicarrier backscatter modulation, and ambient backscatter modulation.

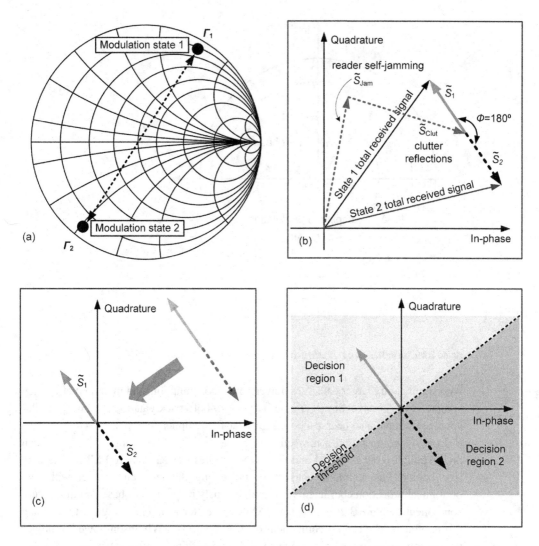

Figure 3.7 (a) Example of PSK modulation at the transponder. (b) Received baseband constellation with arbitrary phase rotation and DC offset. (c) Centered baseband constellation after DC offset removal. (c) BPSK demodulation and symbol mapping.

3.5.1   Quadrature Backscatter Modulation

The analysis presented in the previous section for two backscatter modulation states can be generalized to an arbitrary number of modulation states (M) described by reflection coefficients Γ_m, $m = 1, 2, \ldots, M$. Each scattered vector generated by these modulation states can be detected at the reader using a quadrature demodulation receiver like that in Figure 3.1, which is available in most modern RFID readers. This allows us to perform backscatter data transmissions using quadrature amplitude modulation (QAM) techniques like those used in conventional wireless systems. QAM backscatter [18, 19] enables multi-bit-per-symbol backscatter communications

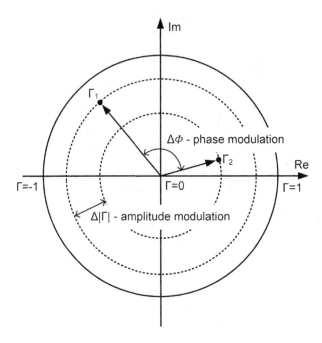

Figure 3.8 Constellation example of mixed amplitude and phase modulation.

with significantly larger data rates, lower power consumption, and increased efficiency compared to conventional single-bit-per-symbol backscatter schemes. Similar approaches have also been proposed for near-field applications [20].

In [19], we proposed a new quadrature backscatter modulation scheme that can produce high-order M-QAM and achieve data rates greater than 100 Mb/s. The core idea behind this approach consists of presenting the incoming carrier with two orthogonal reflection coefficients to simultaneously reflect its in-phase and quadrature components. Figure 3.9(a) shows a QAM-backscatter front end designed to evaluate this approach. The central component of this system is a Wilkinson power combiner whose combining ports are terminated with two varying resistances (R_I and R_Q) that are used to encode the baseband data. To realize orthogonal backscatter modulation, a 45° phase shift is introduced between the two combining ports, which produces an effective phase shift of 90° in the scattered signal.

The in-phase and quadrature backward-propagating waves at the outputs of the combiner circuit are given as in (3.27) for a given incoming wave at the antenna. The reflection coefficient at the antenna port (3.29) can be obtained from the total backward propagating wave at the antenna port (3.28). Note that the in-phase and quadrature components of the reflection coefficient in (3.29) can be controlled independently via two termination impedances, allowing two degrees of freedom for the generation of arbitrary scattering constellations:

$$V_I^- = \frac{V_{ant}^+}{\sqrt{2}}\Gamma_I, \qquad V_Q^- = \frac{V_{ant}^+ e^{j\pi/2}}{\sqrt{2}}\Gamma_Q, \tag{3.27}$$

$$V_{ant}^{-} = \frac{V_I^{-} + V_Q^{-}}{\sqrt{2}} = \frac{V_{ant}^{+}}{\sqrt{2}}\Gamma_I + \frac{V_{ant}^{+}e^{j\pi/2}}{\sqrt{2}}\Gamma_Q, \tag{3.28}$$

$$\Gamma_{ant} = \frac{V_{ant}^{-}}{V_{ant}^{+}} = \frac{\Gamma_I}{2} + j\frac{\Gamma_Q}{2}, \tag{3.29}$$

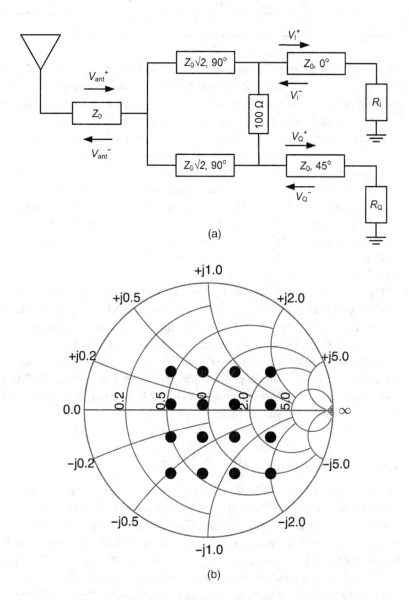

(a)

(b)

Figure 3.9 (a) Diagram of the proposed QAM backscatter circuit. The key component of this system is a Wilkinson power combiner with 45 degree phase shift between the combining arms, which allows for independent modulation of the in-phase and quadrature components of the backscattered signal. (b) Simulated results of a 16-QAM backscatter constellation produced by the proposed front end.

Table 3.2 Parameters of 4-QAM backscatter.

R_I	R_Q	Γ_I	Γ_Q	Γ_{ant}
0	0	-1	-1	$(-1-j)/2$
0	∞	-1	1	$(-1+j)/2$
∞	0	1	-1	$(1-j)/2$
∞	∞	1	1	$(1+j)/2$

where the in-phase and quadrature components are given by $\Gamma_I = (R_I - Z_0)/(R_I + Z_0)$ and $\Gamma_Q = (R_Q - Z_0)/(R_Q + Z_0)$, respectively. For example, a 4-QAM constellation can be produced by loading the backscatter front end using a combination of open and short terminations as in Table 3.2.

Figure 3.9(b) presents simulation results of a 16-QAM backscatter front end. A backscatter modulator front end operating at 2.45 GHz has also been designed and prototyped (see Figure 3.10(a)). In this implementation, each combining arm of the Wilkinson power combiner was terminated with a MOSFET transistor providing a path to ground with an impedance that can be modulated via a bias voltage applied to the MOSFET gate. The backscatter modulator front end was driven with a 2.45 GHz carrier created by an RF signal generator, and an arbitrary waveform generator was used to generate the baseband data to modulate the gates of the in-phase and quadrature MOSFETs.

The 16-QAM signal created by the backscatter modulator was sampled via a directional coupler (used to separate the generated QAM signal from the 2.45 GHz carrier signal) and was then demodulated using a vector signal analyzer. The system performance was evaluated for different data rates and signal-to-noise ratio (SNR) values. Figure 3.10(c) shows the constellation of a 16-QAM signal demodulated by the signal analyzer. In this experiment, an error vector magnitude (EVM) of 8.19% was obtained for a 4 Mb/s data rate and 21.74 dB SNR. For further details, the interested reader is referred to [19].

3.5.2   Spread-Spectrum Backscatter Modulation

The chirp spread spectrum (CSS) modulation technique, originally proposed for radar applications, has recently been applied to digital communications because of its outstanding coverage range and reliability [21–23]. To achieve reliable long-range communications while maintaining reduced power, complexity, and cost of the hardware, backscatter-based CSS modulation approaches have been proposed [22–24].

CSS modulation requires continuously varying the frequency of a sinusoidal signal as a function of time. As the instantaneous frequency corresponds to the rate of change of phase, a CSS chirp signal can be created by modulating the signal phase. A quadrature backscatter modulator system like the one introduced in the previous section (Figure 3.10(a)) was used to produce chirp signals compliant with the LoRa protocol [21] via backscatter modulation [24].

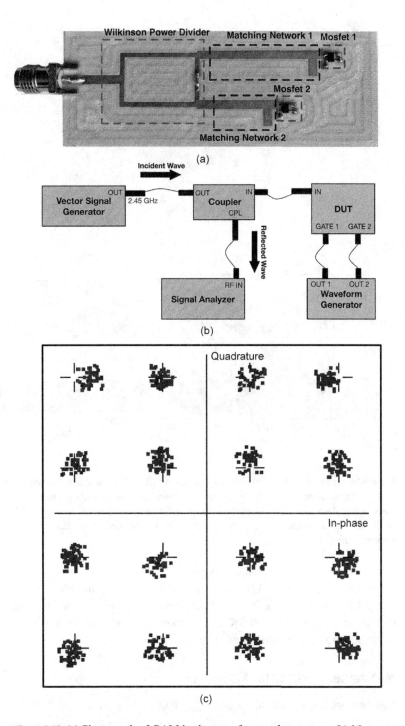

Figure 3.10 (a) Photograph of QAM backscatter front end prototype. (b) Measurement testbed. (c) Measured 16-QAM constellation. Reprinted from [19].

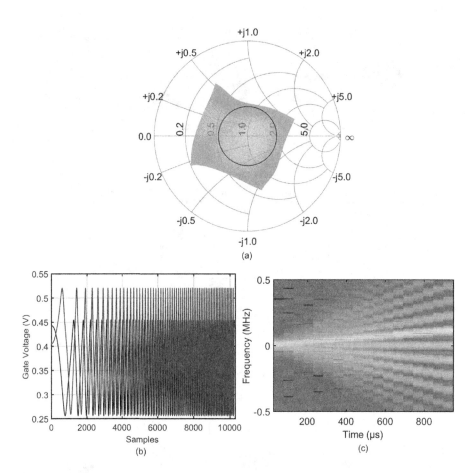

Figure 3.11 (a) Measured reflection coefficient constellation at the output of the quadrature backscatter modulator for various modulation states (gray area) and constant-voltage standing wave ratio (black line). (b) IQ modulating waveforms applied to the in-phase and quadrature inputs of the backscatter modulator. (c) Chirp signal produced by the backscatter modulator. Reprinted from [24].

Figure 3.11 illustrates a chirp signal synthesis performed in [24]. The backscatter constellation in Figure 3.11(a) was obtained by varying the phase of the reflection coefficient while keeping its magnitude constant. This resembles traditional quadrature phase-shift keying (QPSK) modulation. Algorithms and MATLAB code for implementing these modulations are given in Chapter 5.

3.5.3   Multicarrier Backscatter Modulation

Recent work has shown that signals like chaotic, ultra-wide band (UWB), and multi-carrier waveforms can improve the efficiency of wireless power transfer systems when compared to continuous-wave (CW) signals [25–30]. In Chapter 10, we review the literature on multicarrier wireless power transfer and discuss its application for enhancing performance of UHF RFID transponders.

In addition to its potential benefit in terms of energy transfer efficiency, the multi-carrier approach can enable the implementation of enhanced backscatter communication schemes. For example, frequency diversity in backscattered multicarrier signals can be used to counteract multipath fading [26] and correct demodulation errors [33]. Similar concepts propose the use of ancillary power carrier transmitters [31, 32]. Multicarriers can also enable improved medium access control (MAC) schemes for backscatter systems, for example, using frequency division multiple access (FDMA) [33–36].

Figure 3.12 illustrates the multicarrier backscatter communication scheme described in the following sections. In Chapter 7, we experimentally evaluate a similar approach and assess the gain obtained in off-the-shelf UHF RFID transponders using a custom-designed SDR RFID reader with multicarrier capability.

3.5.3.1 Generating a Downlink Modulated Multicarrier

During reader-to-transponder data transmissions, RFID readers modulate a CW carrier signal like that in (3.30) with a baseband message signal $m(t)$. A convenient way to create an RFID modulated multicarrier waveform is to mix this modulated CW signal with a complex baseband multisine envelope like the one in (3.31) having spectral components at $\omega_n = \pm n\Delta\omega$ with $n = 0, 1, \ldots, N$ to yield a transmit bandpass multicarrier like that in (3.32), which has frequency components at $\omega_m = \pm \omega_{\text{carrier}} \pm m\Delta\omega$ with $m = 0, 1, \ldots, 2N + 1$:

$$S_{\text{carrier}}(t) = \text{Re}\left\{A_{\text{carrier}}e^{j\omega_{\text{carrier}}t}\right\}, \tag{3.30}$$

$$S_{\text{env}}(t) = \text{Re}\left\{\sum_{n=0}^{N} A_n e^{j(n\Delta\omega t + \varphi_n)}\right\}, \tag{3.31}$$

$$S_{\text{tx}}(t) = \text{Re}\left\{m(t)\sum_{n=-N}^{N} A_n A_{\text{carrier}}e^{j[(\omega_{\text{carrier}}+n\Delta\omega)t+\varphi_n]}\right\}, \tag{3.32}$$

where N is the number of baseband subcarriers, A_{carrier} and ω_{carrier} are the amplitude and frequency of the reader carrier, respectively, A_n (A_{-n}) and φ_n (φ_{-n}) are the amplitude and phase of the nth baseband subcarrier, respectively, and Δ_ω is the frequency spacing between them. The DC component $(n = 0)$ in the baseband multisine (3.31) yields a spectral component in the bandpass signal (3.32) at the carrier frequency. As shown later, this component is important if the multicarrier signal backscattered by a transponder is to be demodulated by a conventional reader receiver.

For example, a two-tone baseband multisine $(N = 2)$ with a DC component yields a five-tone bandpass multicarrier like the one illustrated in Figure 3.12(b), which has spectral components at $(-\omega_{\text{carrier}} - 2\Delta\omega)$, $(-\omega_{\text{carrier}} - \Delta\omega)$, $-\omega_{\text{carrier}}$, $(-\omega_{\text{carrier}} + \Delta\omega)$, $(-\omega_{\text{carrier}} + 2\Delta\omega)$, $(\omega_{\text{carrier}} - 2\Delta\omega)$, $(\omega_{\text{carrier}} - \Delta\omega)$, ω_{carrier}, $(\omega_{\text{carrier}} + \Delta\omega)$, and $(\omega_{\text{carrier}} + 2\Delta\omega)$. To maximize peak-to-average power ratio (PAPR) in the multicarrier signal and improve its effectiveness in driving transponder energy-harvesting circuits, a

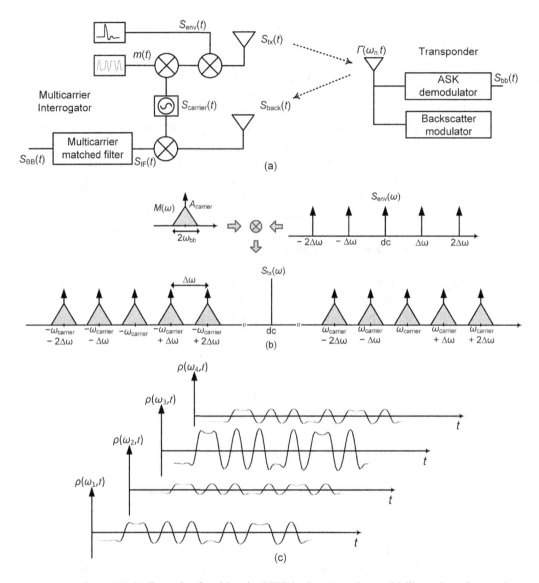

Figure 3.12 (a) Example of multicarrier RFID backscatter scheme. (b) Illustration of transmit-modulated multicarrier resulting from mixing a modulated RFID signal at frequency $\omega_{carrier}$ and bandwidth ω_{bb} with a multisine baseband envelope with subcarrier frequency spacing of $\Delta\omega$. (c) Illustration of time–frequency dependence of modulated backscatter multicarrier.

constant phase progression should be used among the subcarriers [26, 29]. Additionally, to avoid aliasing between the reader-modulated subcarriers and allow a conventional ASK demodulator to retrieve the reader message in the transponder, the frequency spacing between the subcarriers should be at least twice the bandwidth of the reader baseband message, that is, $\Delta\omega \geq 2\omega_{bb}$ (see Figure 3.12(b)).

3.5.3.2 Downlink: Joint Power and Data Transfer via Multicarriers

In the proposed approach, both power and information are transferred from the reader to the transponder via a multicarrier waveform. Multicarrier wireless power transfer is discussed in Chapter 10. To analyze the multicarrier downlink data transmission, assume a noiseless, lossless channel with direct line of sight, and consider that the rectifying circuit of the transponder demodulator is represented by a simple second-order nonlinearity. The demodulator baseband output in response to the multicarrier (3.32) results from the mixture of the symmetric frequency components $(\omega_{carrier} \pm n\Delta\omega)$ and $-(\omega_{carrier} \pm n\Delta\omega)$ with $n = 0, 1, 2, \ldots, N$. For the five-tone multicarrier example $(N = 2)$, the baseband signal retrieved by the transponder demodulator is [26]

$$S_{bb}(t) = \frac{KA_{carrier}^2}{8}[m(t)]^2 \left(A_0^2 + 2A_1^2 + 2A_2^2\right), \tag{3.33}$$

where K is the second-order coefficient of the transponder demodulator nonlinearity. As can be seen, the transponder demodulator output is proportional to the square of the reader message signal (for binary ASK modulation, $m(t)$ and $[m(t)]^2$ carry the same information). Therefore, for proper multicarrier design (e.g., as in Figure 3.12), a conventional transponder ASK demodulator can retrieve the reader message without requiring any modification to the transponder hardware. Transponder architectures optimized for multicarrier operation have also been proposed in the literature [37–39].

3.5.3.3 Uplink: Multicarrier Backscatter

Backscatter communications are typically implemented in a sequential manner where the reader illuminates the transponder with an unmodulated carrier, and the transponder reflects a portion of this carrier by presenting a modulated complex reflection coefficient to the incoming signal that encodes its message. To study the multicarrier backscatter mechanism, we assume that the reader illuminates the transponder with a multicarrier of the form (3.32) with the modulating baseband signal $m(t)$ held constant. The reflection coefficient presented to the multicarrier by the transponder antenna is a function of frequency and time,

$$\Gamma(\omega_k, t) = \rho(\omega_k, t)e^{j\theta(\omega_k, t)}, \tag{3.34}$$

where $\rho(\omega_k, t)$ and $\theta(\omega_k, t)$ are the frequency- and time-dependent amplitude and phase of the reflection coefficient presented to the kth subcarrier at frequency ω_k, respectively. The complex reflection coefficient at each frequency also depends on the unmodulated frequency response of the transponder antenna and modulator. Assuming the reader radiates an unmodulated multicarrier like that in (3.32) with $m(t) = 1$ through a fading channel, the transponder backscattered signal intercepted by the reader is given by

$$S_{back}(t) = \mathrm{Re}\left\{\sum_{n=-N}^{N}\sum_{m=0}^{M-1}\Gamma(\omega_n, t)A_n A_{carrier}\alpha_{n,m}e^{j\left(n\Delta\omega t + \varphi_{n,m}\right)}e^{-j\omega_n \tau_{n,m}}e^{j\omega_{carrier}t}\right\}, \tag{3.35}$$

where $\alpha_{n,m}$, $\varphi_{n,m}$, and $\tau_{n,m}$ are the roundtrip attenuation, phase shift, and delay of the nth subcarrier received through the mth path, respectively, and M is the number of signal paths. For simplicity, here we assume that the transponder produces symmetrical modulation around the reader carrier. But in practice asymmetric modulation of the sideband subcarriers may introduce distortion in the transponder baseband message signal (see Chapter 7 for more details). The first step in demodulating the transponder message backscattered on the multicarrier is to correlate the received signal (3.35) with the reader local oscillator carrier, which yields the complex baseband envelope,

$$\widetilde{S_{BB}}(t) = S_{BB}^I(t) + jS_{BB}^Q(t) = \sum_{n=0}^{N}\sum_{m=0}^{M-1}\Gamma(\omega_n,t)A_n A_{carrier}\alpha_{n,m}e^{j(n\Delta\omega t + \phi_{n,m})}, \qquad (3.36)$$

where $\phi_{n,m} = \varphi_{n,m} - \omega_n\tau_{n,m}$ is the round-trip phase shift of the nth subcarrier at the mth path. The in-phase and quadrature components of the baseband envelope are given by

$$S_{BB}^I(t) = \sum_{n=0}^{N}\sum_{m=0}^{M-1}\{\rho(\omega_n,t)\cos[\theta(\omega_n,t)]A_n A_{carrier}\alpha_{n,m}\cos(n\Delta\omega t + \varphi_{n,m})$$
$$- \rho(\omega_n,t)\sin[\theta(\omega_n,t)]A_n A_{carrier}\alpha_{n,m}\sin(n\Delta\omega t + \varphi_{n,m})\},$$

$$S_{BB}^Q(t) = \sum_{n=0}^{N}\sum_{m=0}^{M-1}\{\rho(\omega_n,t)\cos[\theta(\omega_n,t)]A_n A_{carrier}\alpha_{n,m}\sin(n\Delta\omega t + \varphi_{n,m})$$
$$- \rho(\omega_n,t)\sin[\theta(\omega_n,t)]A_n A_{carrier}\alpha_{n,m}\cos(n\Delta\omega t + \varphi_{n,m})\}.$$

Assuming that there is a single line-of-sight path between the reader and transponder with unitary response ($M = 1$, $\alpha_{n,m} = 1$, $\phi_{n,m} = 0$) and that a low-pass filter with cut-off frequency below $\Delta\omega$ is used to isolate the baseband component at DC ($N = 0$), then the transponder baseband component modulated on the center subcarrier simplifies to

$$\widetilde{S_{BB}}(t) = S_{BB}^I(t) + jS_{BB}^Q(t) = \rho(\omega_0,t)A_n A_{carrier}\{\cos[\theta(\omega_n,t)] + j\sin[\theta(\omega_n,t)]\}. \qquad (3.37)$$

This approach presents minimal complexity and makes it possible for conventional reader receivers to demodulate the multicarrier message without any hardware modification [26]. In Chapter 7, we discuss improved approaches based on matched filters and fast Fourier transform (FFT) that can be used to leverage multicarrier frequency diversity.

3.5.4   Ambient Backscatter Modulation

Communication via ambient electromagnetic backscatter has recently been explored in the literature [40–44]. In this technique, an "opportunistic" backscatter radio leverages radio transmissions from existing wireless systems by reflecting their signals to convey information. This approach, which requires no dedicated radio transmitters, has been demonstrated in the TV and FM [40, 41], Wi-Fi [42], and Bluetooth and ZigBee [43, 44] bands.

3.6 Backscatter Power Budget Analysis

To perform a power budget analysis for a typical backscatter UHF RFID system, we consider the scenario presented in Figure 3.13, for which we make the following assumptions:

- The reader employs co-located transmit and receive antennas with gains G_{tx}, G_{rx} respectively.
- The receiver has sensitivity S_{reader} and the transmitter delivers power P_{tx} to the antenna at wavelength λ (frequency $f = c/\lambda$, where c is the speed of light in free space).
- The reader and transponder are in direct line of sight of one another and are separated by a distance d (with $d \gg \lambda$ for far-field operation).

The effective aperture of the antenna along with the Friis formula determines the RF power transferred to the transponder antenna (3.38), and the antenna radar cross section combined with the radar equation determines the power scattered by the transponder and received by the reader (3.39) [3]:

$$P_{tag} = A_e \frac{P_{tx} G_{tx}}{4\pi d^2}, \tag{3.38}$$

$$P_{rx} = \sigma \frac{P_{tx} G_{tx} G_{rx} \lambda^2}{(4\pi)^3 d^4}. \tag{3.39}$$

The aperture and radar cross section of an antenna loaded with an impedance Z_L are given by (3.40) and (3.41), respectively [45, 46]:

$$A_e = \frac{G_{tag} \lambda^2}{4\pi} \left(1 - |\Gamma_L|^2\right), \tag{3.40}$$

$$\sigma = \frac{G_{tag}^2 \lambda^2}{4\pi} |\Gamma_0 - \Gamma_L|^2, \tag{3.41}$$

where G_{tag} is the gain of the transponder antenna, Γ_L is the reflection coefficient as given by (3.16), and Γ_0 accounts for the antenna structural mode scattering [14]. The

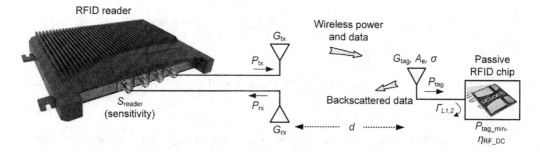

Figure 3.13 Typical backscatter UHF RFID power budget scenario. Note: images of reader and transponder are used for illustration purposes only.

term $\left(1 - |\Gamma_L|^2\right)$ in (3.40) accounts for the mismatch between the transponder antenna and its load impedance. For a total mismatch ($\Gamma_L = 1$) no power is delivered to the transponder chip, and under conjugate impedance matching ($\Gamma_L = 0$) A_e takes the usual form given by the Friis formula.

3.6.1   Average Aperture and Differential Cross Section

Considering again an encoding scheme that switches symmetrically between two modulation states with equal probabilities and reflection coefficients Γ_{L1} and Γ_{L2} leads to an average effective aperture (3.42) and differential radar cross section (3.43) [45] of

$$\Delta A_e = \frac{G_{tag}\lambda^2}{4\pi}\left(1 - \frac{1}{2}\left(|\Gamma_{L1}|^2 + |\Gamma_{L2}|^2\right)\right) = \frac{G_{tag}\lambda^2}{4\pi}(1 - \beta), \tag{3.42}$$

$$\Delta\sigma = \frac{G_{tag}^2\lambda^2}{16\pi}|\Gamma_{L1} - \Gamma_{L2}|^2 = \frac{G_{tag}^2\lambda^2}{4\pi}\rho, \tag{3.43}$$

where we have defined

$$\beta = \frac{1}{2}\left(|\Gamma_{L1}|^2 + |\Gamma_{L2}|^2\right), \qquad \rho = \frac{1}{4}|\Gamma_{L1} - \Gamma_{L2}|^2,$$

For improved accuracy, transponder gain degradation due to attachment to objects and line-of-sight blockage should also be considered. Note that the common-mode term Γ_0 in the radar cross section of the individual modulation states (3.41) does not contain information and cancels out in the differential case (3.43).

3.6.2   Communication Range Limitations

Replacing the antenna aperture in (3.38) and radar cross section in (3.39) with (3.42) and (3.43), respectively, and considering polarization mismatch and multipath fading, yields the power collected by the transponder (3.44) and the power received by the reader (3.45):

$$P_{tag}[\text{dBm}] = P_{tx} + G_{tx} + G_{tag} + 20\log_{10}\left(\frac{\lambda}{4\pi d}\right) + 10\log_{10}(1 - \beta) + \theta_f - F_f, \tag{3.44}$$

$$P_{rx}[\text{dBm}] = P_{tx} + G_{tx} + G_{rx} + 2G_{tag} + 40\log_{10}\left(\frac{\lambda}{4\pi d}\right) + 10\log_{10}(\rho) + \theta_f + \theta_b - F_f, \tag{3.45}$$

where θ_f and θ_b describe the polarization mismatch between the transponder antenna and the reader transmit and receive antennas, respectively, and F_f is the forward fade margin, required in a faded channel to achieve the same transponder read rate that

Table 3.3 Typical parameters used in the model of Figure 3.14.

Parameter	Description	Value
P_{tx}	Reader transmitted power	30 dBm
G_{tx}	Reader transmitter antenna gain	6 dBi (circular)
G_{rx}	Reader receiver antenna gain	6 dBi (circular)
S_{reader}	Reader receiver sensitivity	−80 dBm
$P_{tag_{min}}$	Transponder activation power level	−15 dBm
G_{tag}	Transponder antenna gain	2 dBi (linear)
θ_f	Reader to transponder polarization mismatch	−3 dB (circular to linear)
θ_b	Transponder to reader polarization mismatch	−3 dB (linear to circular)
F_f	Forward fade margin	0 dB (no multipath)
f_c	RF carrier frequency	866 MHz
λ	Wavelength	0.35 m
Γ_1	Transponder reflection state 1	0 (absorptive state)
Γ_2	Transponder reflection state 2	−1 (reflective state)
d	Reader to transponder distance	1 to 100 m

would be achieved in a flat channel. The results of (3.44) and (3.45) are presented in Figure 3.14 as a function of the distance between the reader and transponder for the typical parameters given in Table 3.3.

The main conclusions from this analysis are twofold. First, in particular, while a reader receiver sensitivity of −80 dBm would allow a coverage range of about 20 m, the maximum coverage range is cut short at 6 m because of the transponder's large minimum activation power level (−15 dBm). Second, typically, the most stringent limitation in the communication range of passive backscatter systems is imposed by the forward wireless power transfer link rather than the reader receiver sensitivity. This result, which contrasts with conventional wireless communication systems (primarily limited by noise and receiver sensitivity) holds true for state-of-the-art transponders with activation power levels as low as −20 dBm. This has motivated the investigation of several techniques to improve the forward link in passive-backscatter systems, including those discussed in Section 3.7 and Chapter 10.

3.7 Wireless Power Transfer in Backscatter Systems

The total wireless power transfer efficiency of a passive-backscatter system relates the DC power delivered to the transponder load to the DC power consumed by the reader transmitter (Figure 3.15). This efficiency has three contributions: transmitter DC to RF efficiency, RF channel beam efficiency, and transponder RF to DC efficiency:

$$\eta_{\text{end-to-end}} = \frac{P_{\text{DCload}}}{P_{\text{supply}}} = \eta_{\text{DC-RF}}\eta_{\text{Beam}}\eta_{\text{RF-DC}} = \frac{P_{\text{RFtx}}}{P_{\text{supply}}}\frac{P_{\text{RFrx}}}{P_{\text{RFtx}}}\frac{P_{\text{DCload}}}{P_{\text{RFrx}}} \qquad (3.46)$$

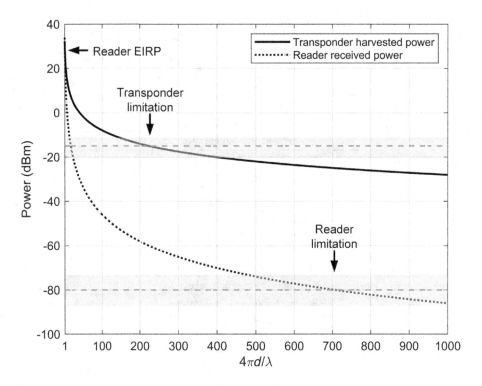

Figure 3.14 Typical passive-backscatter power budget.

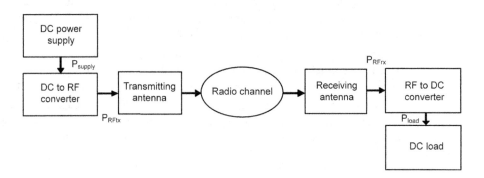

Figure 3.15 Overall wireless power transmission efficiency.

We review various approaches reported in the literature for enhancing RF beam efficiency and RF to DC efficiency (a summary is presented in Table 3.4). DC to RF efficiency is typically dominated by transmit power amplifier efficiency and is outside the scope of this discussion.

Table 3.4 Summary of techniques used to improve wireless power transfer efficiency in passive-backscatter systems.

	Approach	Comment	References
Waveform design	Multicarrier waveforms UWB pulses Chaotic signals White noise Pulsed-CW signals	Waveform design optimization for improved efficiency in energy harvesting circuits.	[25–30]
Beam efficiency improvement	Antenna beamforming and beam-steering	Achieving high gain and wide coverage areas simultaneously.	[47–52]
	Antenna polarization diversity	Using circularly and multi-polarized reader antennas to improve sensitivity to misalignment between reader and transponder antennas.	[53, 56]
	Multi-receiving antennas	Collecting power from vertical and horizontal incoming fields using orthogonal transponder antennas.	[57–61]
	Frequency, space, and antenna diversity	Use of diversity schemes like frequency-hopping spread spectrum in the reader to combat multipath fading.	[62–65]
Conversion efficiency improvement	Rectifier topology optimization	Optimizing the efficiency of energy harvesters at the circuit and system levels.	[66–73]
	Device threshold compensation	Improving the sensitivity of CMOS-based harvesting circuits to small power levels via built-in threshold compensation.	[74–76]
Other approaches	Ancillary transponder DC power supply Ancillary RF power supply Synergistic energy harvesting	Use of ancillary DC power supply and energy harvesting mechanisms (e.g., solar, thermal) in the transponder, and ancillary wireless power transmitters to boost the power available to transponders.	[17, 31, 32, 59, 81–83]

3.7.1   Improving RF Beam Efficiency

The RF beam efficiency, which relates the RF power at the transponder antenna to the RF power injected into the reader transmit antenna, depends on various factors including the polarization mismatch between transmitter and receiver antennas, line-of-sight blockage, path loss, transponder detuning, and multipath fading. This component [the first logarithmic term in (3.44)] is the major bottleneck in non-directive wireless power transfer approaches like those typically used in passive-backscatter RFID. For example, an 866 MHz carrier signal undergoes a path loss in free space of at least 50 dB over a 10 m distance.

3.7.1.1   Antenna Beamforming and Beam-Steering

Point-to-point microwave wireless power transmission links usually employ highly directive antennas with very narrow beams and gains as high as 20 dBi [47, 48]. RFID reader systems, on the other hand, use low- to medium-gain antennas (typically 5–7 dBi) as they are required to scan wide areas. Some techniques used in conventional wireless systems such as antenna beamforming and beam-steering have been proposed to achieve high gain and wide coverage areas in passive RFID applications [49–52]. Besides providing increased antenna gain over wide areas, these techniques enable advanced applications such as high-resolution RFID localization [52]. Notably, antenna beamforming/beam-steering is a cornerstone of 5G and the next generations of wireless communications systems, which promise unprecedented communication bandwidths and a new plethora of applications and services.

3.7.1.2   Antenna Polarization Diversity

Another way to enhance beam efficiency is to use antenna polarization diversity [53–56]. Passive RFID transponders typically employ linearly polarized antennas to lower their cost and complexity and are usually attached to objects that can appear in the interrogation field of the reader in arbitrary orientations. To mitigate polarization misalignment, RFID readers are usually equipped with circularly polarized antennas. Circular-to-linear polarization arrangements are less sensitive to the transponder antenna orientation and are more robust in harsh communication environments such as vehicular communications compared to linear-to-linear polarization arrangements [53, 54]. This is achieved at the expense of reduced gain.

To leverage the benefits of linear and circular polarizations, multi-linear polarization schemes have been explored. For instance, the Speedway reader portal from Impinj Inc. uses a dual-linear phased array antenna scheme to deliver the power and coverage range of a linear antenna with the ability to read transponders with arbitrary orientations.

3.7.1.3   Multi-Receiving Antennas

By orthogonally arranging two linearly polarized receiving antennas, one can collect power from both the vertical and horizontal components of an incoming electrical field and improve the system sensitivity to the polarization angle. Multi-antenna schemes have been proposed to mitigate polarization misalignment in wireless power transmission and passive RFID applications [57–61].

3.7.1.4   Multipath Fading Mitigation

Due to multipath fading in UHF RFID channels [62], transponders can be rendered powerless at locations where the incoming scattered waves combine destructively, that is, at null field locations or fades. Based on the observation that a subtle change in the radio communication channel can dramatically alter the multipath fading pattern and cancel a fade, several techniques based on frequency, space, and antenna diversity have been proposed to mitigate multipath fading. Off-the-shelf UHF RFID readers commonly use a frequency diversity technique called frequency-hopping spread spectrum [63]. Antenna polarization diversity has also been proposed to combat multipath fading in RFID applications [64, 65].

3.7.2 Improving RF–DC Conversion Efficiency

Schottky- and CMOS-based rectifying devices used in passive transponders typically present poor conversion efficiency at the low-input-power regimes where transponders often operate when far away from the reader. Optimizing the efficiency of energy harvesting circuits is therefore key to improving the communication range and overall performance of passive-backscatter systems.

The research on this topic spans multiple areas, including the study of new rectifying devices and topologies [66–70], harmonic rectenna design [71–73], and rectifying threshold compensation techniques [74–76]. Figure 3.16 summarizes the results of reported energy harvesting efficiencies in several ISM bands [55, 56, 77–80]. The following conclusions can be drawn from this survey:

- In general, the conversion efficiency increases with the power level applied to the energy harvesting circuit and decreases with the operating frequency. The latter trend is due to increased parasitic losses at higher frequencies while the former is because of the non-zero turn-on voltage of the rectifying devices.

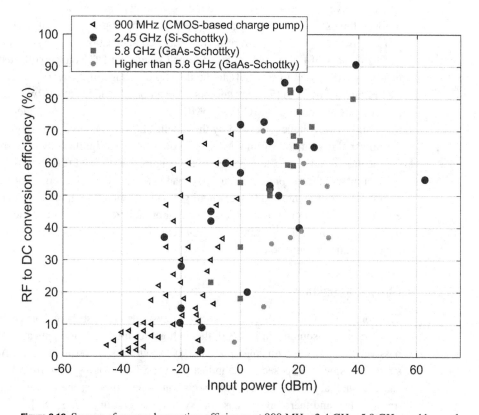

Figure 3.16 Survey of energy harvesting efficiency at 900 MHz, 2.4 GHz, 5.8 GHz, and beyond.

- Most energy harvesting circuits used in passive UHF RFID are based on CMOS diode-connected transistor technology, operate at relatively low power levels (typically below 0 dBm), and present, in general, low power conversion efficiencies.
- Energy harvesting circuits for directed microwave wireless power transmission (typically at 2.4 GHz, 5.8 GHz, and beyond) are mostly based on discrete Schottky diodes, operate at significantly higher input power levels, and present high efficiencies (for example, one of the most efficient energy harvesting circuits ever reported was a GaAs-Pt Schottky diode rectenna, which recorded 90.6% efficiency at 2.45 GHz for an input power of 39 dBm [80]).
- Ambient electromagnetic energy harvesting circuits typically operate at very low power regimes (below -30 dBm) and present poor power conversion efficiencies in those regimes.

3.7.2.1   CMOS Threshold Compensation

CMOS diode-connected devices are widely used in rectifying circuits for passive UHF RFID transponders because of their compatibility with standard fabrication processes. But CMOS-based rectifiers typically present limited efficiencies due to increased transistor turn-on voltages. This is especially important at the low power regimes in which passive transponders often operate. Several approaches have been proposed to mitigate this issue, including improved rectifying topologies, physical device parameter optimization, and device threshold voltage compensation [74–76].

The authors of [74] proposed an approach where the DC voltage collected by the rectifier itself is used to compensate the threshold voltage of the rectifying transistors (see Figure 3.17(a)). In [75], the authors inserted photovoltaic cells between the gates and sources or drains of nMOS and pMOS transistors operating as diode-connected transistors. The voltage generated by these cells allowed the circuit to compensate for the threshold voltages of the MOSFET devices and significantly improve their rectification efficiency at low power levels (Figure 3.17(b) and (c)).

A similar concept presented in [76] made use of thermal energy harvesting to compensate for the threshold of a Schottky diode and therefore improve its conversion efficiency at low input power levels. The authors utilized a thermo-electric generator to harvest thermal energy and used the collected DC voltage as an external bias source for their Schottky junction.

3.7.3   Other Approaches

The communication range of passive RFID systems can also be enhanced by using ancillary power sources. In [31, 32], the authors doubled the communication range of passive transponders by using an auxiliary CW wireless power transmitter. A similar strategy has been proposed in the industry – the STAR 3000 system from Mojix Inc. can use ancillary wireless power transmission nodes to boost the power available to transponders and therefore increase the system's operational range [81].

In [82], the authors proposed the use of solar energy harvesting to improve the operation of a passive RFID transponder. They used a solar cell to power an external RF oscillator and then combined the generated signal with the signal received from the

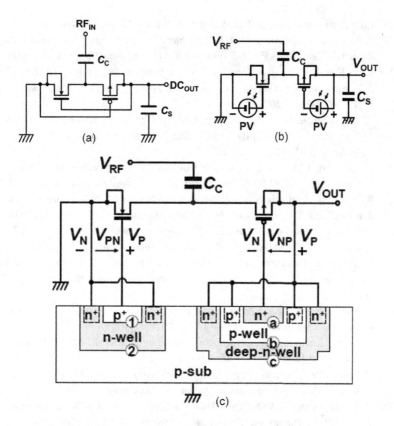

Figure 3.17 CMOS threshold-compensation. (a) Self-threshold cancellation. (b) Photovoltaic-assisted rectifier. (c) Detail of photovoltaic rectifier shown in (b). These voltage doubler circuits use diode-connected nMOS and pMOS transistors. Reprinted from [74, 75].

reader at the transponder's antenna port. Under sufficient light incidence, this scheme can allow for substantially improved transponder operational range.

Some off-the-shelf transponder devices incorporate ancillary DC and antenna ports that enable the implementation of improved antenna diversity [59] and battery-assisted [17, 83] schemes. For example, the UCODE G2iL chip from NXP provides an input DC port that allows the implementation of a battery-assisted transponder. According to the manufacturer, a minimum transponder activation power as low as -27 dBm can be achieved with this device when using an external 1.85 V battery [17].

References

[1] H. Stockman, "Communication by Means of Reflected Power," *Proceedings of the Institute of Radio Engineers*, 36(10):1196–1204, 1948.

[2] D. B. Harris, "Radio Transmission Systems with Modulatable Passive Responder," U.S. patent number 2927321, 1960.

[3] J. D. Kraus, *Antennas*, 2nd ed. McGraw-Hill Book Company, New York, 1988.

[4] G. Marrocco, "The Art of UHF RFID Antenna Design: Impedance-Matching and Size-Reduction Techniques," *IEEE Antennas and Propagation Magazine*, 50(1):66–79, 2008.

[5] K. V. S. Rao, P. V. Nikitin, and S. F. Lam, "Antenna Design for UHF RFID Tags: A Review and a Practical Application," *IEEE Transactions on Antennas and Propagation*, 53(12):3870–3876, 2005.

[6] P. V. Nikitin, K. V. S. Rao, R. Martinez, and S. F. Lam, "Sensitivity and Impedance Measurements of UHF RFID Chips," *IEEE Transactions on Microwave Theory and Techniques*, 57(5):1297–1302, 2009.

[7] R. Kronberger, A. Geissler, and B. Friedmann, "New Methods to Determine the Impedance of UHF RFID Chips," in *IEEE International Conference on RFID*, 2010.

[8] R. E. Collin, "Limitations of the Thevenin and Norton Equivalent Circuits for a Receiving Antenna," *IEEE Antennas and Propagation Magazine*, 45(2):119–124, 2003.

[9] J. B. Andersen and R. G. Vaughan, "Transmitting, Receiving, and Scattering Properties of Antennas," *IEEE Antennas and Propagation Magazine*, 45(4):93-98, 2003.

[10] A. W. Love, "Comment on 'Limitations of the Thevenin and Norton Equivalent Circuits for a Receiving Antenna'," *IEEE Antennas and Propagation Magazine*, 45(4):98–99, 2003.

[11] D. Pozar, "Scattered and Absorbed Powers in Receiving Antennas," *IEEE Antennas and Propagation Magazine*, 46(1):144–145, 2004.

[12] W. K. Kahn and H. Kurss, "Minimum-Scattering Antennas," *IEEE Transactions on Antennas and Propagation*, 13(5):671–675, 1965.

[13] P. Rogers, "Application of the Minimum Scattering Antenna Theory to Mismatched Antennas," *IEEE Transactions on Antennas and Propagation*, 34(10):1223–1228, 1986.

[14] P. V. Nikitin and K. V. S. Rao, "Theory and Measurement of Backscattering from RFID Tags," *IEEE Antennas and Propagation Magazine*, 48(6):212–218, 2006.

[15] K. Finkenzeller, *RFID Handbook: Fundamentals and Applications in Contactless Smart Cards, Radio Frequency Identification and Near-Field Communication*, 3rd ed. Wiley, Chichester, 2010.

[16] D. M. Dobkin, *The RF in RFID: Passive UHF in Practice*, 2nd ed. Newnes, Oxford, 2012.

[17] NXP, "UCODE G2iL and G2iL+ Product Ddatasheets." [Online], available at: www.nxp.com

[18] S. J. Thomas, E. Wheeler, J. Teizer, and M. S. Reynolds, "Quadrature Amplitude Modulated Backscatter in Passive and Semipassive UHF RFID Systems," *IEEE Transactions on Microwave Theory and Techniques*, 60(4):1175–1182, 2012.

[19] R. Correia, A. Soares Boaventura, and N. Borges Carvalho, "Quadrature Amplitude Backscatter Modulator for Passive Wireless Sensors in IoT Applications," *IEEE Transactions on Microwave Theory and Techniques*, 65(4):1103–1110, 2017.

[20] J. Besnoff, M. Abbasi, and D. S. Ricketts, "High Data-Rate Communication in Near-Field RFID and Wireless Power Using Higher Order Modulation," *IEEE Transactions on Microwave Theory and Techniques*, 64(2):401–413, 2016.

[21] Semtech, "LoRa Modulation Basics." [Online], available at: https://semtech.my.salesforce.com/sfc/p/#E0000000JelG/a/2R00000010Ju/xvKUc5w9yjG1q5Pb2IIkpolW54YYqGb.frOZ7HQBcRc

[22] V. Talla et al., "LoRa Backscatter: Enabling The Vision of Ubiquitous Connectivity," *Proceedings of the ACM on Interactive, Mobile, Wearable and Ubiquitous Technologies*, 1(3):105, 2017.

[23] Y. Peng et al., "PLoRa: A Passive Long-Range Data Network from Ambient LoRa Transmissions," in *Proceedings of the 2018 Conference of the ACM Special Interest Group on Data Communication (SIGCOMM '18)*. ACM Press, New York, 2018, pp. 147–160.

[24] R. Correia et al., "Chirp Based Backscatter Modulation," in *2019 IEEE MTT-S International Microwave Symposium (IMS)*, 2019, pp. 279–282.

[25] M. S. Trotter and G. D. Durgin, "Survey of Range Improvement of Commercial RFID Tags with Power Optimized Waveforms," in *2010 IEEE International Conference on RFID*, 2010, pp. 195–202.

[26] A. J. Soares Boaventura and N. Borges Carvalho, "Extending Reading Range of Commercial RFID Readers," *IEEE Transactions on Microwave Theory and Techniques*, 61(1):633–640, 2013.

[27] A. Collado and A. Georgiadis, "Improving Wireless Power Transmission Efficiency Using Chaotic Waveforms," in *International Microwave Symposium*, 2012.

[28] C.-C. Lo, et al., "Novel Wireless Impulsive Power Transmission to Enhance the Conversion Efficiency for Low Input Power," in *IEEE MTT-S International Microwave Workshop Series on Innovative Wireless Power Transmission*, 2011, pp. 55–58.

[29] A. J. Soares Boaventura, A. Collado, A. Georgiadis, and N. Borges Carvalho, "Spatial Power Combining of Multi-Sine Signals for Wireless Power Transmission Applications," *IEEE Transactions on Microwave Theory and Techniques*, 62(4):1022–1030, 2014.

[30] N. Pan, et al., "Multi-Sine Wireless Power Transfer with a Realistic Channel and Rectifier Model," in *2017 IEEE Wireless Power Transfer Conference (WPTC)*, 2017, pp. 1–4.

[31] J. S. Park et al., "Extending the Interrogation Range of a Passive UHF RFID System with an External Continuous Wave Transmitter," *IEEE Transactions on Instrumentation and Measurement*, 58(8):2191–1297, 2010.

[32] H.-c. Liu, Y.-t. Chen, and W.-s. Tzeng, "A Multi-Carrier UHF Passive RFID System," in *Proceedings of the International Symposium on Applications and Internet Workshops (SAINTW)*, 2007, p. 21.

[33] M. Fischer et al., "An Experimental Study on the Feasibility of a Frequency Diverse UHF RFID System," *IEEE Access*, 7:132311–132323, 2019.

[34] A. Bletsas, S. Siachalou, and J. N. Sahalos, "Anti-Collision Backscatter Sensor Networks," *IEEE Transactions on Wireless Communications*, 8(10):5018–5029, 2009.

[35] N. Rajoria et al., "Multi-Carrier Backscatter Communication System for Concurrent Wireless and Batteryless Sensing," in *2017 International Conference on Wireless Communications, Signal Processing and Networking (WiSPNET)*, 2017, pp. 1078–1082.

[36] D. Belo et al., "A Selective, Tracking, and Power Adaptive Far-Field Wireless Power Transfer System," *IEEE Transactions on Microwave Theory and Techniques*, 67 (9):3856–3866, 2019.

[37] H.-C. Liu and C.-R. Kuo, "Reader Coverage Analysis for Multi-Carrier Passive UHF RFID Systems," *Wireless Personal Communications*, 59(1):123–133, 2011.

[38] M.-H. Lee, C.-Y. Yao, and H.-C. Liu, "Passive Tag for Multi-Carrier RFID Systems," in *2011 IEEE International Conference on Parallel and Distributed Systems*, 2011, pp. 872–876.

[39] M.-H. Lee et al., "Demodulator for Multi-Carrier UHF RFID Tags," in *2013 IEEE International Symposium on Consumer Electronics (ISCE)*, 2013, pp. 5–6.

[40] V. Liu et al., "Ambient Backscatter: Wireless Communication Out of Thin Air," in *Proceedings of the ACM SIGCOMM Conference*, 2013, pp. 39–50.

[41] S. N. Daskalakis et al., "Ambient Backscatterers Using FM Broadcasting for Low Cost and Low Power Wireless Applications," *IEEE Transactions on Microwave Theory and Techniques*, 65(12):5251–5262, 2017.

[42] B. Kellogg et al., "Wi-Fi Backscatter: Internet Connectivity for RF-Powered Devices," *ACM SIGCOMM Computer Communication Review*, 44(4): 607–618, 2014.

[43] J. F. Ensworth and M. S. Reynolds, "Every Smart Phone Is a Backscatter Reader: Modulated Backscatter Compatibility with Bluetooth 4.0 Low Energy (BLE) Devices," in *2015 IEEE International Conference on RFID*, 2015, pp. 78–85.

[44] V. Iyer et al., "Inter-Technology Backscatter: Towards Internet Connectivity for Implanted Devices," in *Proceedings of the 2016 ACM SIGCOMM Conference (SIGCOMM'16)*. ACM, New York, pp. 356–369.

[45] P. Pursula, "Analysis and Design of UHF and Millimetre Wave Radio Frequency Identification," Ph.D. thesis, Faculty of Information and Natural Sciences, Helsinki University of Technology, Finland, 2009.

[46] J. D. Griffin and G. D. Durgin, "Complete Link Budgets for Backscatter-Radio and RFID Systems," *IEEE Antennas and Propagation Magazine*, 51(2):11–25, 2009.

[47] Z. Popovic et al., "Lunar Wireless Power Transfer Feasibility Study," Technical Report DOE/NV/25946-488, 2008.

[48] N. Shinohara, "Beam Control Technologies With a High-Efficiency Phased Array for Microwave Power Transmission in Japan," *Proceedings of the IEEE*, 101(6):1448–1463, 2013.

[49] M. Abbak and I. Tekin, "RFID Coverage Extension Using Microstrip-Patch Antenna Array," *IEEE Antennas and Propagation Magazine*, 51(1):185–191, 2009.

[50] A. M. Ndifon et al., "Performance Improvements of Multicast RFID Systems Using Phased Array Antennas and Phase Diversity," in *2017 IEEE International Conference on RFID Technology & Application (RFID-TA)*, 2017, pp. 51–56.

[51] R. Sadr, "RFID Beam Forming System," U.S. patent 8768248, 2014.

[52] G. P. Eloy, "Steerable Phase Array Antenna RFID Tag Locater and Tracking System and Methods," U.S. patent 8742896B2, 2014.

[53] D. Megnet and H. Mathis, "Circular Polarized Patch Antenna for 5.8 GHz Dedicated Short-Range Communication," in *Proceedings of the 39th European Microwave Conference*, 2009, pp. 638–641.

[54] B. Lua and P. Li, "Design of Circular Polarization Microstrip Antenna in RFID Reader for 5.8 GHz Electronic Toll Collection Application," in *International Conference on Microwave Technology and Computational Electromagnetics*, 2009, pp. 84–87.

[55] J. O. McSpadden and K. Chang, "A Dual Polarized Circular Patch Rectifying Antenna at 2.45 GHz for Microwave Power Conversion and Detection," in *IEEE MTT-S International Microwave Symposium Digest*, vol. 3, 1994, pp. 1749–1752.

[56] T. Paing et al., "Wirelessly-Powered Wireless Sensor Platform," in *European Conference on Wireless Technologies*, 2007, pp. 241–244.

[57] D. W. Duan and D. J. Friedman, "Cascaded DC Voltages of Multiple Antenna RF Tag Front-End Circuits," U.S. patent 6243013, 2001.

[58] J. Hyde, O. Onen, and R. A. Oliver, "RFID Tags Combining Signals Received from Multiple RF Ports," U.S. patent 20060049917, 2006.

[59] Impinj, "Monza X-2K Dura Datasheet." [Online], available at: www.impinj.com

[60] P. V. Nikitin and K. V. S. Rao, "Performance of RFID Tags with Multiple RF Ports," in *IEEE Antennas and Propagation Society International Symposium*, 2007, pp. 5459–5462.

[61] J. D. Griffin, "High-Frequency Modulated-Backscatter Communication Using Multiple Antennas," Ph.D. thesis, Georgia Institute of Technology, 2009.

[62] J. Mitsubi, "UHF Band RFID Readability and Fading Measurements in Practical Propagation Environment," Auto-ID Lab Japan, White Paper Series, Edition 1.

[63] Alien Technology, "Reader Interface Guide," 2018. [Online], available at: www.alientechnology.com

[64] J. D. Griffin and G. D. Durgin, "Multipath Fading Measurements for Multi-Antenna Backscatter RFID at 5.8 GHz," in *IEEE International Conference on RFID*, 2009, pp. 322–329.

[65] D. H. Ha, J.-S. Ahn, and Y.-W. Choe, "A Multi-Polarized Antenna System for Combating Multipath Fading in UHF/RFID Frequency Band in Moving Vehicles," in *2010 International Conference on Communications and Mobile Computing (CMC)*, 2010, pp. 63–67.

[66] C.-Y. Liou et al., "High-Power and High-Efficiency RF Rectifiers Using Series and Parallel Power-Dividing Networks and Their Applications to Wirelessly Powered Devices," *IEEE Transactions on Microwave Theory and Techniques*, 61(1):616–624, 2013.

[67] C. Gómez, J. A. García, A. Mediavilla, and A. Tazón, "A High Efficiency Rectenna Element using E-pHEMT Technology," *12th GaAs Applications Symposium*, 2004.

[68] Q. Li, J. Wang, Y. Han, and H. Min, "Design and Fabrication of Schottky Diode in Standard CMOS Process," *Chinese Journal of Semiconductors*, 26(2):238–242, 2005.

[69] W. C. Brown, "Rectenna Technology Program: Ultra Light 2.45 GHz Rectenna and 20 GHz Rectenna," Raytheon Company Technical Report, 1987.

[70] H. Sun, Z. Zhong, and Y.-X. Guo, "An Adaptive Reconfigurable Rectifier for Wireless Power Transmission," *IEEE Microwave and Wireless Components Letters*, 23 (9):492–494, 2013.

[71] Z. Harouni, L. Osman, and A. Gharsallah, "Efficient 2.45 GHz Rectenna Design with High Harmonic Rejection for Wireless Power Transmission," *International Journal of Computer Science Issues*, 7(5):424–427, 2010.

[72] S. Imai et al., "Efficiency and Harmonics Generation in Microwave to DC Conversion Circuits of Half-Wave and Full-Wave Rectifier Types," in *2011 IEEE MTT-S International Microwave Workshop Series on Innovative Wireless Power Transmission: Technologies, Systems, and Applications*, 2011, pp. 15–18.

[73] K. Hatano et al., "Development of Class-F Load Rectennas," *2011 IEEE MTT-S International Microwave Workshop Series on Innovative Wireless Power Transmission: Technologies, Systems, and Applications*, 2011, pp. 251–254.

[74] K. Kotani and T. Ito, "High Efficiency CMOS Rectifier Circuit with Self-Vth-Cancellation and Power Regulation Functions for UHF RFIDs," in *2007 IEEE Asian Solid-State Circuits Conference*, 2007, pp. 119–122.

[75] K. Kotani, "Highly Efficient CMOS Rectifier Assisted by Symmetric and Voltage-Boost PV-Cell Structures for Synergistic Ambient Energy Harvesting," in *IEEE Custom Integrated Circuits Conference*, 2013, pp. 1–4.

[76] M. Virili et al., "Performance Improvement of Rectifiers for WPT Exploiting Thermal Energy Harvesting," *Wireless Power Transfer*, 2(1):22–31, 2015.

[77] Y.-J. Ren, M.-Y. Li, and K. Chang, "35 GHz Rectifying Antenna for Wireless Power Transmission," *Electronics Letters*, 43(11): 602–603, 2007.

[78] E. Falkenstein, M. Roberg, and Z. Popovic, "Low-Power Wireless Power Delivery," *IEEE Transactions on Microwave Theory and Techniques*, 60(7):2277–2286, 2012.

[79] C. R. Valenta and G. D. Durgin, "Harvesting Wireless Power: Survey of Energy-Harvester Conversion Efficiency in Far-Field, Wireless Power Transfer Systems," *IEEE Microwave Magazine*, 15(4):108–120, 2014.

[80] J. O. McSpadden, L. Fan, and K. Chang, "Design and Experiments of a High-Conversion-Efficiency 5.8-GHz Rectenna," *IEEE Transactions on Microwave Theory and Techniques*, 46(12):2053–2060, 1998.

[81] Mojix, "STAR 3000 System." [Online], available at: www.mojix.com

[82] A. Georgiadis and A. Collado, "Improving Range of Passive RFID Tags Utilizing Energy Harvesting and High Efficiency Class-E Oscillators," *Sixth European Conference on Antennas and Propagation*, 2012, pp. 3455–3458.

[83] A.P. Sample et al., "Photovoltaic Enhanced UHF RFID Tag Antennas for Dual Purpose Energy Harvesting," in *2011 IEEE International Conference on RFID*, 2011, pp. 146–153.

4 An Overview of the ISO 18000-63 Standard

For the experiments conducted in this book we selected ISO 18000-63/EPC Gen2V2 [1, 2], which is currently the most prevalent standard for RFID in the UHF band. This chapter gives an overview of key aspects of this standard relating to coding and modulation, protocol timing, data security, transponder singulation, and memory access. This discussion is not intended to be exhaustive, and the reader is referred to [1, 2] for further details.

4.1 Reader-to-Transponder Encoding and Modulation

Reader-to-transponder communications under ISO 18000-63 use pulse interval encoding (PIE). This encoding scheme is designed to provide transponders with uninterrupted wireless power transfer during reader data transmissions. In PIE, each binary data symbol is encoded with a power-off pulse following a full-power interval of at least the same duration as the power-off interval (see Figure 4.1(a)). This prevents the interruption of energy transfer to the transponder for certain combinations of transmitted symbols, for example long strings of on–off-keying-modulated '0' symbols. In ISO 18000-63, each reader *Query* command is preceded by a *preamble* frame (Figure 4.1(b)) which also indicates the beginning of an inventory round, and all other commands begin with a *Frame sync* sequence (Figure 4.1(c)).

ISO 18000-63 defines three modulation formats for reader-to-transponder communications: double-sideband amplitude-shift keying (DSB-ASK), single-sideband amplitude-shift keying (SSB-ASK), and phase-reversal amplitude-shift keying (PR-ASK). SSB-ASK and PR-ASK provide superior bandwidth efficiency compared to DSB-ASK. A low-cost and low-complexity approach for implementing PR-ASK modulation is presented in Chapter 6.

Figure 4.2(a) shows the ISO 18000-63 transmit mask for a dense-interrogator environment. The standard also specifies a mask for a multiple-interrogator environment [1, 2]. ISO 18000-63 also specifies the transmit RF envelope, which determines the characteristics of the transmit spectrum (see Figure 4.2(b)). To limit transmit spectrum occupancy and comply with regulated transmit masks, baseband pulse shaping is typically required in the transmitter (this is detailed in Chapter 5). In addition to the ISO 18000-63 requirements, readers must comply with mask specifications imposed by ETSI in Europe [3] and the FCC in the USA [4].

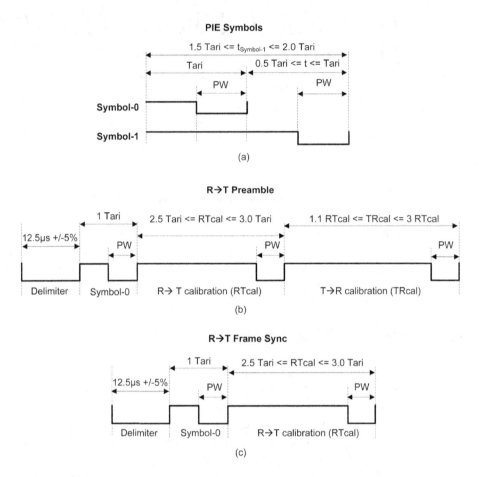

Figure 4.1 Reader-to-transponder signaling. (a) PIE code. (b) PIE Preamble. (c) PIE Frame synchronization.

4.2   Transponder-to-Reader Encoding and Modulation

ISO 18000-63 specifies two encoding and modulation schemes for transponder-to-reader communications. Transponders can encode data using FM0 (Figure 4.3(a)) or Miller-modulated subcarrier codes and can modulate the RF carrier using either ASK or PSK. The modulation method is selected by the transponder whereas the encoding scheme is chosen by the reader through the *Query* command that initiates each inventory round. Both the modulation and encoding types must remain unchanged for the entire inventory round. For the experiments in this book, we used FM0 encoding. For information on Miller-modulated subcarrier encoding, refer to [1, 2].

The FM0 encoding scheme possesses key features that simplify the design of the RFID reader. First, by encoding information in the signal transitions, this scheme allows for more robust backscatter communication and easier synchronization, clock recovery, and data decoding at the reader. FM0 inverts its phase at the boundaries of

Table 4.1 RF envelope parameters.

Tari	Parameter	Symbol	Min.	Nom.	Max.	Units
	Modulation depth	$(A - B)/A$	80	90	100	%
6.25 µs	RF envelope ripple	$M_h = M_I$	0		$0.05(A - B)$	V/m, A/m
to	RF envelope rise time	$t_{r,10-90\%}$	0		0.33Tari	µs
25 µs	RF envelope fall time	$t_{f,10-90\%}$	0		0.33Tari	µs
	RF pulse width	PW	max (0.265Tari, 2)		0.525Tari	µs

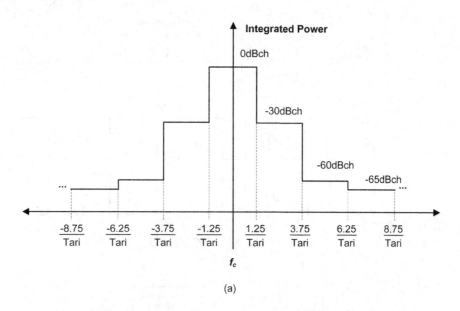

(a)

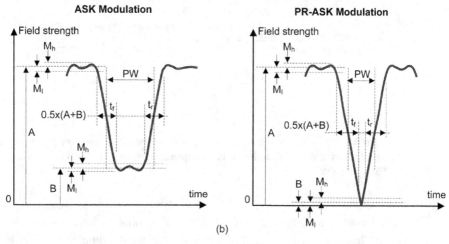

(b)

Figure 4.2 (a) ISO 18000-63 transmit mask for dense-interrogator environment. Channels are defined with respect to the *Tari* parameter defined in Figure 4.1. (b) Time-domain RF envelope specifications (see Table 4.1).

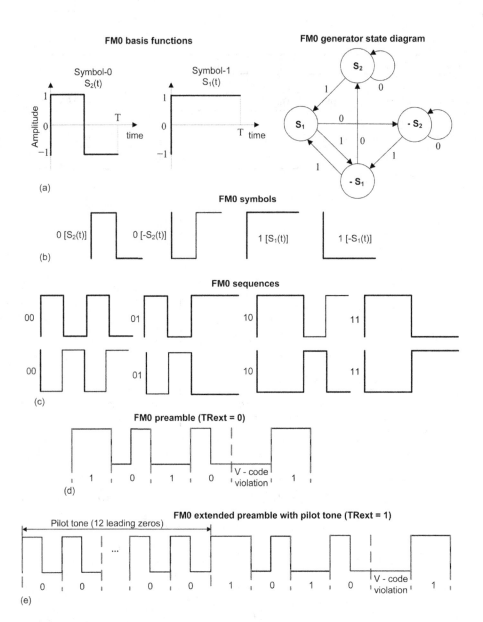

Figure 4.3 Transponder-to-reader signaling. (a) FM0 basis functions and generator state diagram. (b) FM0 symbols. (c) Example of FM0 sequences. (d) Regular FM0 preamble. (e) Extended FM0 preamble.

symbols 0 and 1, and in the middle of symbol 0 (see Figure 4.3(a) and (b)). In Chapter 6, we leverage this feature to implement a simple yet robust FM0 decoder using standard built-in peripherals of our MCU. Furthermore, we exploit the FM0 memory feature to detect errors in the message received from the transponder even before cyclic redundancy check (CRC) verification.

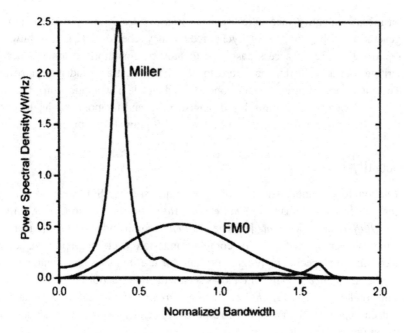

Figure 4.4 Spectrum of FM0 and Miller-modulated subcarrier codes (reprinted from [5]).

Second, FM0 and Miller-modulated subcarrier codes present minimal DC components (see Figure 4.4). This helps to mitigate self-jamming and DC offset problems
encountered in direct-conversion receiver (DCR) architectures commonly used in
passive UHF RFID readers (this aspect is discussed in detail in Chapter 8). Another
key feature of the FM0 encoding scheme is the use of orthogonal basis functions (see
Figure 4.3(b)), which enables the implementation of enhanced correlation-based
decoders. This feature is further explored in the next chapter.

Depending on the reader instruction, the transponder can begin its transmission
with either a regular FM0 preamble (Figure 4.3(d)) or an extended FM0 preamble
(Figure 4.3(e)). The extended FM0 preamble prepends a pilot tone composed of 12
FM0 zeros to the message. The pilot tone is used to assist in clock recovery and can
help to mitigate baseband transients in DCR receivers (see Chapter 8), but it introduces communication overhead and suppressing it can allow a speed-up in processing.
To improve preamble uniqueness, a symbol (V) that violates the FM0 boundary
transition rule is introduced in the preamble pattern (Figure 4.3(d) and (e)).

4.3 New ISO 18000-63 Memory and Security Features

In this section, we give an overview of security and memory enhancements introduced
by the ISO 18000-63 standard and its counterpart, EPC Gen2V2, the ratified versions
of ISO 18000-6C and EPC Gen2. Traditionally, ISO 18000-6C and Gen2 support

basic data integrity mechanics such as transponder 16-bit pseudorandom number generation, 5-bit and 16-bit cyclic redundancy check (CRC), and bitwise XOR. Optionally, a 32-bit access password to prevent unauthorized transponder memory writing and a 32-bit kill password to deactivate the transponder are also available. To address new security/privacy concerns and enable new applications, ISO 18000-63 and EPC Gen2v2 have introduced new security and memory management features, made available through a set of 12 additional (optional) commands [1, 2].

4.3.1 Security

To allow for the implementation of cryptographic suites, ISO 18000-63 introduced six new security commands: *Challenge*, *Authenticate*, *SecureComm*, *AuthComm*, *ReadBuffer*, and *KeyUpdate* [2]. The *Challenge* command allows the reader to instruct a transponder population to precompute and store a cryptographic value or values according to a specific cryptographic suite for use in a subsequent authentication.

The *Authenticate* command allows secure data transactions between the reader and transponder, while the *AuthComm* and *SecureComm* commands allow authenticated and encrypted data transactions, respectively, and typically encapsulate other protected reader commands in their message fields.

The *KeyUpdate* command allows a reader to write or overwrite a cryptographic key in a transponder after an authentication has been performed. The *ReadBuffer* command allows the reader to access results of cryptographic operations stored in the transponder's response buffer. For more information on these security features, the interested reader is referred to [6, 7].

4.3.2 Privacy

ISO 18000-63 also introduces a privacy command, *Untraceable*, which allows access to the transponder to be restricted. For instance, this command allows a reader with an asserted *Untraceable* privilege to instruct the transponder to deny memory access to readers without appropriate privileges. A convenient feature implemented by the *Untraceable* command makes it possible to restrict access to the transponder at the physical level by instructing it to reduce its operating range. The new features also make it possible to temporarily hide parts of the transponder memory, including words of the EPC.

4.3.3 Memory

The ISO 18000-63 update introduced a file system for the transponder user memory bank and four new commands for managing that file system: *FileOpen*, *FileList*, *FileSetup*, and *FilePrivilege*. Like its preceding version, ISO 18000-63 divides the transponder memory into four banks (Bank00, reserved memory; Bank01, EPC memory; Bank10, tag identifier (TID) memory; and Bank11, user memory). But ISO 18000-63 adds file partitioning to the user memory Bank11 and allows the user

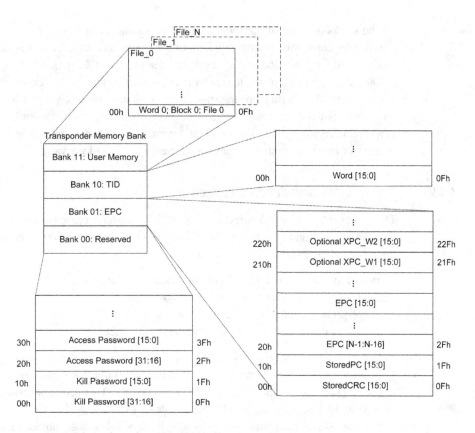

Figure 4.5 Configuration of an ISO 18000-63 transponder memory.

to customize the type, size, and access privileges of each file (see Figure 4.5). This opens new possibilities such as the storage of data owned by different parties on the same transponder with different access privileges, or the storage of product lifecycle and maintenance information.

4.4   Transponder Singulation and Memory Access

This section presents an overview of transponder identification and memory access, including aspects of transponder population selection, collision resolution, and singulation, according to ISO 18000-63.

4.4.1   Transponder Population Selection

Upon entering the field of an RFID reader, transponders power up, enter their ready state, and wait for instructions from the reader. The reader must singulate a transponder in the transponder population for further transactions. This process starts with the reader broadcasting a *Select* command whose parameters define the selection criteria for the sub-population of transponders for the subsequent inventory round. If a single transponder enters the interrogation field of the reader, as exemplified in Section 4.6, it automatically matches the selection criteria (see [1, 2] for details).

During the selection phase, the reader may issue an (optional) *Challenge* command to cryptographically challenge the transponder population. Upon receiving a *Challenge*, the transponder pre-computes and stores a cryptographic value or values in its buffer according to a specific cryptographic suite for use in a subsequent authentication. Transponders can execute a *Challenge* command from any state except killed (deactivated).

4.4.2   Transponder Singulation and Access

The transponder singulation process under ISO 18000-63 is based on a random-slotted collision arbitration algorithm (*Q*-algorithm). Upon receiving a *Query* command from the reader, each transponder,

1) randomly picks a time slot to communicate with the reader;
2) loads the selected slot number into its slot counter;
3) decrements its slot counter per the reader's request;
4) replies to the reader once its slot counter expires (reaches zero).

Transponder singulation is typically implemented using the commands *Query*, *QueryRep*, *QueryAdjust*, *Ack*, and *NAK*. The reader starts the singulation process by issuing a *Query* command containing the *Q* parameter. This parameter, which ranges from 0 to 15, specifies the available number of slots $(N = 2^Q)$. Upon receiving a *Query* command, each transponder randomly chooses a slot number in the range $[0, 2^Q - 1]$. Transponders that select slot number zero immediately backscatter a random 16-bit number, RN16, in the interval $[0, 2^{16} - 1]$.

If a single transponder picks slot 0 (no collision), then the reader issues an *Ack* command containing the received RN16 to acknowledge that transponder (if two or more transponders pick slot 0, a collision will occur). Upon reception of an *Ack* with a matching RN16, the transponder replies with its EPC data.

The inventory round continues with the reader issuing one or more *QueryAdjust* or *QueryRep* commands. Each *QueryRep* command causes the transponders participating in the inventory round to decrement their slot counter by 1. To readjust the *Q* parameter without changing other round parameters, the reader uses the *QueryAdjust* command. This command may instruct the transponders to increment, decrement, or keep their *Q*. Upon receiving a *QueryAdjust* command, a transponder first updates its *Q* then chooses a slot number. If a transponder reaches zero because of a *QueryAdjust* or *QueryRep*, the singulation process continues similarly to the initial *Query*.

After finishing inventorying the transponder population, the reader may initiate a singulated transponder into the access state by issuing a *Req_RN* command that contains that transponder's RN16. Upon receiving a *Req_RN* with a matching RN16, the transponder responds with a new 16-bit random number or Handle. This Handle is to be used by the reader in subsequent memory accesses. Figure 4.6 depicts a simplified diagram of the transponder singulation and access after ISO 18000-63.

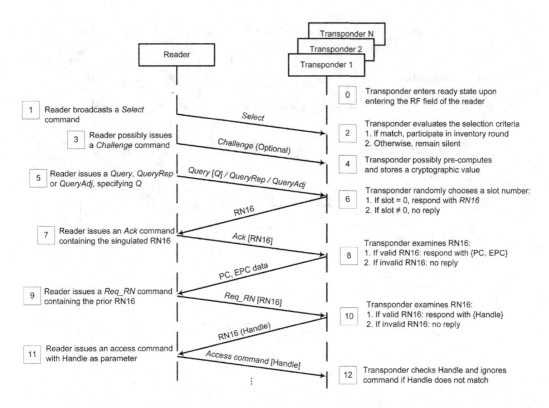

Figure 4.6 Simplified diagram of transponder singulation, inventory, and access.

4.4.3   Colliding Transponder Responses

If two or more transponders choose slot number zero at the same time, then a collision will occur as they reply simultaneously to the reader with their RN16. In this case, the reader may try to detect the collision at the waveform level and resolve an RN16 from one of the transponders, or not resolve the collision and issue a *QueryAdjust*, *QueryRep*, or *NAK*.

4.4.4   Optimal *Q*-Slot Selection

While a small Q value may lead to faster inventory time for small transponder populations, it can increase the collision probability for large transponder populations. Conversely, a large Q unnecessarily increases the inventory time for small transponder populations and renders the singulation process inefficient. For optimal performance, Q should be adjusted dynamically based on the number of transponders in the interrogation zone, which can be estimated based, for instance, on the number of empty slots, collided slots, and single response slots. A dynamic algorithm for setting Q is presented in [2].

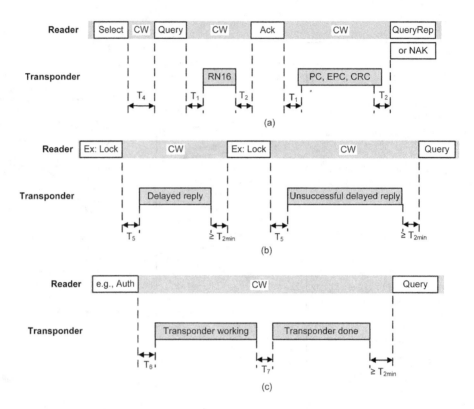

Figure 4.7 Protocol timing (see Table 4.2 for parameter details). (a) Immediate reply. (b) Delayed reply. (c) In-process reply.

For the experiments conducted in this book, we considered a single transponder in the reader interrogation field; by setting Q to zero (single slot), we caused the first transponder entering the interrogation field to be immediately singulated.

4.4.5   Communication Timing

Because cryptographic algorithms require significant computation time, ISO 18000-63 introduces flexible protocol timing. In addition to the conventional immediate transponder reply (Figure 4.7(a)), the standard specifies a delayed reply (Figure 4.7(b)) and an in-process reply (Figure 4.7(c)) for transponders that require a significant amount of time to complete their operations. Here, we outline the main ISO 18000-63 timing parameters.

The time between a transponder response and the next reader command is defined by T_2 (Figure 4.7(a)). If a transponder in the reply or acknowledged state does not hear from the reader before a maximum period $T_{2\,max}$, it times out. The reader response timing $T_{2\,max}$ becomes more restrictive as the backscatter link frequency (BLF)[1]

---

[1] BLF denotes either the FM0 symbol rate or the Miller subcarrier frequency.

Table 4.2 Timing for the examples of Figure 4.7. Refer to [1, 2] for more details.

Parameter	Minimum	Maximum	Description
T_1	$\max(\text{RTcal}, 10T_{\text{pri}}) \times (1 - |\text{FrT}|) - 2$ μs	$\max(\text{RTcal}, 10T_{\text{pri}}) \times (1 + |\text{FrT}|) + 2$ μs	Immediate reply time from reader command to transponder reply
T_2	$3.0T_{\text{pri}}$	$20.0T_{\text{pri}}$	Interrogator reply time if a transponder is to demodulate the reader signal
T_4	2.0RTcal		Minimum time between reader commands
T_5	$\max(\text{RTcal}, 10T_{\text{pri}}) \times (1 - |\text{FrT}|) - 2$ μs	20 ms	Delayed reply time from reader command to transponder reply
T_6	$\max(\text{RTcal}, 10T_{\text{pri}}) \times (1 - |\text{FrT}|) - 2$ μs	20 ms	In-process reply time from reader command to the first transponder reply
T_7	$\max(250$ μs$, T_{2\max})$	20 ms	In-process reply time between transponder replies

Notes:

1. T_{pri} denotes either the commanded period of an FM0 symbol or the commanded period of a single Miller subcarrier cycle, as appropriate.
2. The maximum value for T_2 applies only to transponders in the reply or acknowledged states. In all other states, T_2 is unrestricted. A transponder is allowed a tolerance of $20.0T_{\text{pri}} \leq T_{2\max} \leq 32T_{\text{pri}}$ in determining whether T_2 has expired.
3. A reader may transmit a new command prior to interval T_2 (i.e., during a transponder response). In this case, the transponder may ignore the new command.
4. FrT is the frequency tolerance [1, 2].

increases. This can be problematic for readers with limited processing speed. In Chapter 7, we discuss some approaches to mitigate this problem.

The time between a reader command and the transponder response depends on the type of transponder response being used. This time is denoted T_1 for immediate replies (Figure 4.7(a)), T_5 for delayed replies (Figure 4.7(b)), and T_6 or T_7 for in-process replies (Figure 4.7(c)).

The minimum amount of time the reader should wait before issuing a new command after receiving the transponder reply is defined by $T_{2\min} = 3T_{pri}$, and the minimum time between consecutive reader commands is defined by $T_4 = 2\text{RTcal}$ (see Figure 4.7(a)).

A transponder using a delayed reply can spend up to 20 ms processing a command (Figure 4.7(b)), while a transponder using an in-process reply can spend longer than that but is required to beacon a busy signal at least once every 20 ms to notify the reader that the command is still being processed (Figure 4.7(c)).

4.5   The ISO 18000-63 Command Set

Table 4.3 lists the main commands specified by ISO 18000-63. The standard also reserves command codes for the implementation of custom and battery-assisted tag (BAP) operations. See Tables 4.4–4.17 for typical examples of reader-to-transponder transactions. Note that the examples presented are for illustration purposes only; different settings may be required for specific cases. The CRC checksums are computed according to Table 4.18.

Table 4.3 ISO 18000-63 main command set.

Command	Binary code	Length (bits)	Group	Mandatory (Yes/No)	Protection mechanism
Select	1010	> 44	Select	Yes	CRC-16
Challenge	11010100	> 48	Select	No	CRC-16
Query	1000	22	Inventory	Yes	CRC-5
QueryAdjust	1001	9	Inventory	Yes	Unique length
QueryRep	00	4	Inventory	Yes	Unique length
ACK	01	18	Inventory	Yes	Unique length
NAK	11000000	8	Inventory	Yes	Unique length
Req_RN	11000001	40	Access memory	Yes	CRC-16

Table 4.3 (*cont.*)

Command	Binary code	Length (bits)	Group	Mandatory (Yes/No)	Protection mechanism
Read	11000010	> 57	Access memory	Yes	CRC-16
Write	11000011	> 58	Access memory	Yes	CRC-16
Kill	11000100	59	Access memory	Yes	CRC-16
Lock	11000101	60	Access memory	Yes	CRC-16
Access	11000110	56	Access memory	No	CRC-16
BlockWrite	11000111	> 57	Access memory	No	CRC-16
BlockErase	11001000	> 57	Access memory	No	CRC-16
BlockPermalock	11001001	> 66	Access memory	No	CRC-16
Authenticate	11010101	> 64	Access security	No	CRC-16
AuthComm	11010111	> 42	Access security	No	CRC-16
SecureComm	11010110	> 56	Access security	No	CRC-16
KeyUpdate	11100010 00000010	> 72	Access security	No	CRC-16
TagPrivilege	11100010 00000011	78	Access memory	No	CRC-16
ReadBuffer	11010010	67	Access security	No	CRC-16
FileOpen	11010011	52	Access file	No	CRC-16
FileList	11100010 00000001	71	Access file	No	CRC-16
FilePrivilege	11100010 00000100	68	Access file	No	CRC-16
FileSetup	11100010 00000101	71	Access file	No	CRC-16
Untraceable	11100010 00000000	62	Access privacy	No	CRC-16

4.6   Reader–Transponder Transactions

Table 4.4 Reader *Select* command.

Parameter	Command	Target	Action	MemBank	Pointer	Length	Mask	Truncate	CRC-16
Example	1010	100	000	01	00100000	00000000	00000000	0	1010001000000110
Description	*Select*	Target: SL	Assert matching, deassert non-matching	EPC memory bank (01)	EPC starting address	0-length mask – tag automatically matches	Mask value	Disable truncation	16-bit CRC

Table 4.5 Reader *Query* command.

Parameter	Command	DR	M	TRext	Sel	Session	Target	Q	CRC-5
Example	1000	0	00	1	00	00	0	0000	10011
Description	*Query*	Divide ratio=8	FM0 coding ($M = 1$)	Use pilot tone	All tags respond Ignore SL flag	Session (S0)	Inventory flag (A)	$Q = 0$, immediate reply	5-bit CRC

Table 4.6 Reader *QueryRep* command (if no transponder answered a *Query*, one or more *QueryRep* commands are issued).

Parameter	Command	Session
Example	00	00
Description	*QueryRep*	Session (S0)

Table 4.7 Transponder RN16 response (singulated).

Parameter	RN16*
Example	0011000011000111
Description	16-bit random number

Table 4.8 Reader *Ack* command.

Parameter	Command	RN16
Example	01	0011000011000111
Description	*Ack*	16-bit random number (previously received from transponder)

Table 4.9 Transponder EPC response.

Parameter	PC	EPC*	CRC-16
Example	0011000000000000	E2003411B802011111365086 (hexadecimal)	0110000101011101
Description	16-bit PC	96-bit EPC data	16-bit CRC

* Transponder data collected with the reader presented in Chapter 6.

Table 4.10 Reader *NAK* command.

Parameter	Command
Example	11000000
Description	*NAK*

Table 4.11 Reader *Req_RN* command.

Parameter	Command	RN16	CRC-16
Example	11000001	0011000011000111	1000111010011010
Description	*Req_RN*	Prior 16-bit random number received from transponder	16-bit CRC

Table 4.12 Transponder RN16 response (Handle).

Parameter	RN16 (new RN16 called Handle)	CRC-16
Example	0000011110000111	1001101000001000
Description	New 16-bit random number (Handle)	16-bit CRC

Table 4.13 Reader *Read* command (using Handle as parameter).

Parameter	Command	MemBank	WordPtr	WordCount	RN16	CRC-16
Example	11000010	01	00000010	00000001	0000011110000111	0111111111110011
Description	*Read*	Mem bank (01-EPC)	Starting address pointer (PC word)	Number of words to be read (1)	Prior 16-bit Handle	16-bit CRC

Table 4.14 Transponder response.

Parameter	Header	Memory Words	RN16	CRC-16
Example	0	0011000000000000	0000011110000111	1011001001001111
Description	Header bit (0 – Success)	Read data	Handle	16-bit CRC

Table 4.15 Reader *Write* command (using Handle as parameter).

Parameter	Command	MemBank	WordPtr	Data	RN16	CRC-16
Example	11000011	01	00000010	0011000000000000	0000011110000111	0100000111101100
Description	*Write*	Mem bank (01-EPC)	Starting address pointer (PC word)	RN16 XORed with word to be written	Prior 16-bit Handle	16-bit CRC

Table 4.16 Transponder response.

Parameter	Header	RN16	CRC-16
Example	0	0000011110000111	1011110100011001
Description	Header bit (0 – Success)	Handle	16-bit CRC

Table 4.17 Reader *Challenge* command.

Parameter	Command	Reserved	IncRepLen	Immed	CSI	Length	Message	CRC-16
Example	11010100	00	1	1	00000000	—	—	—
Description	*Challenge*	Should be set to 00	Include length in reply	Transmit result with EPC	First ISO/IEC 29167 suite	Length of message	Depends on CSI	16-bit CRC

Table 4.18 CRC-16 and CRC-5 definitions.

		CRC-16		
CRC Type	Length	Polynomial	Preset	Residue
ISO/IEC 13239	16 bits	$x^{16} + x^{12} + x^5 + 1$	FFFF$_h$	1D0F$_h$
		CRC-5		
CRC Type	Length	Polynomial	Preset	Residue
—	5 bits	$x^5 + x^3 + 1$	01001_2	00000_2

References

[1] ISO/IEC "18000-63:2013 Information Technology – Radio Frequency Identification for Item Management – Part 63: Parameters for Air Interface Communications at 860 MHz to 960 MHz Type C," 2013.

[2] EPCglobal Inc. "EPC Radio-Frequency Identity Protocols Generation-2 UHF RFID Specification for RFID Air Interface Protocol for Communications at 860 MHz–960 MHz (version 2.0.0 ratified)," 2013.

[3] ETSI "EN 302 208 V3.2.0, Radio Frequency Identification Equipment Operating in the Band 865 MHz to 868 MHz with Power Levels up to 2 W and in the Band 915 MHz to 921 MHz with Power Levels up to 4 W; Harmonised Standard for Access to Radio Spectrum," 2018.

[4] FCC "Title 47, Part 15, Operation within the Bands 902–928 MHz, 2435–2465 MHz, 5785–5815 MHz, 10500–10550 MHz, and 24075–24175 MHz," 2024.

[5] J. Wang, et al., "System Design Considerations of Highly-Integrated UHF RFID Reader Transceiver RF Front-End," in *Ninth International Conference on Solid-State and Integrated-Circuit Technology*, 2008, pp. 1560–1563.

[6] D.W. Engels et al., "On Security with the New Gen2 RFID Security Framework," in *Proceedings of the IEEE International Conference on RFID*, 2013.

[7] H.-Y. Chien, "Efficient Authentication Scheme with Tag-Identity Protection for EPC Class 2 Generation 2 Version 2 Standards," *International Journal of Distributed Sensor Networks*, 13(3), 2017.

5   Digital Signal Processing for RFID Applications

5.1   Introduction

In this chapter, we study key aspects of RFID communication under ISO 18000-63 [1, 2]. To evaluate the reader downlink and uplink digital signal processing blocks in Figure 5.1, we used a combination of custom and native MATLAB functions. To test the implemented scripts and algorithms, we synthesized various custom transponder waveforms.

In this discussion, we focus on the digital baseband processing; we assume that the transponder signal intercepted by the reader has been down-converted to baseband and is available for processing after digitization, and the reader baseband data is up-converted to the desired RF carrier frequency after being converted to the analog domain. Implementations of the RF/analog part are discussed in the following chapters.

The key functions used here are provided in the main text, and evaluation script examples are available online (files *EvaluateReader2TagDownlink.m*, *EvaluateTag2ReaderUplink.m*). The scripts provided were tested in MATLAB R2021a with the communications toolbox. Even though our discussion considers the ISO 18000-63 standard, the scripts and algorithms can be tailored as needed for other protocols and custom designs.

5.2   Transmitter Signal Processing

Here, we discuss the ISO 18000-63 RFID reader transmit chain (top of Figure 5.1) and provide code to evaluate typical reader baseband transmissions with variable parameters. By properly selecting the transmit parameters, the generated waveform can be optimized to meet transmission specifications such as the transmit spectrum mask analyzed at the end of this section. The transmit waveform data produced here can be used in a real RFID reader to implement a lookup table–based waveform generator. This approach, which has been used in commercial RFID reader application-specific integrated circuits (ASICs) [3], was adopted in the SDR reader design presented in Chapter 7. The waveform pre-synthesis approach can also be helpful for research experiments; for example, by simply uploading waveforms into an arbitrary waveform

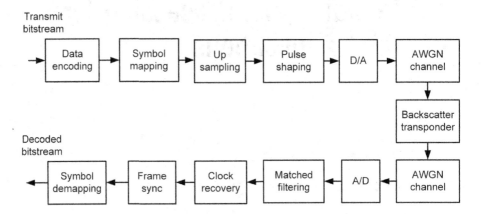

Figure 5.1 Simplified diagram of RFID communication chain including reader transmitter and receiver, RFID transponder, and communication channel. D/A: digital to analog converter; A/D: analog to digital converter.

generator or vector signal generator, one can query transponders without needing a dedicated RFID reader.

5.2.1   Data Encoding

ISO 18000-63 reader-to-transponder data transmissions are encoded using PIE. As mentioned before, this encoding scheme is designed to enable stable wireless power delivery to the transponder during data transmissions, but it presents poor spectral efficiency compared to standard on–off keying (OOK) modulation due to the additional encoding transitions introduced in the signal (see previous chapter). Fortunately, the ISO 18000-63 standard offers efficient modulation schemes, namely single-sideband modulation (Section 5.2.4) and phase-reversal modulation (Section 5.2.5), which allow the overall transmit spectrum efficiency to improve.

In the implementation provided in function *EncodeModulateInterpolate*(), each PIE symbol '0' is encoded using a binary symbol '1' followed by a binary symbol '0', and each PIE symbol '1' contains three consecutive binary '1's followed by a binary '0'. The binary '0' and '1' represent the low and high portions of the PIE symbol, respectively (see Figure 5.2). The resulting encoded binary sequence is then interpolated to the appropriate sampling rate (see next section).

5.2.2   Interpolation and Pulse Shaping

Rectangular pulse shaping (the square-shaped signal in Figure 5.2) is the simplest interpolation method used in communications. Each symbol is confined to one symbol period and does not interfere with neighboring symbols. Assuming no dispersion or

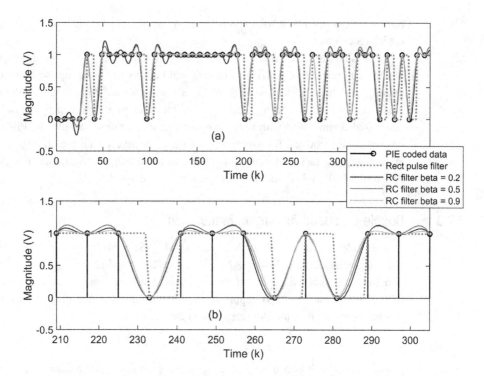

Figure 5.2 Illustration of PIE encoding and interpolation. (a) Time-domain depiction of PIE data interpolated with a rectangular pulse filter and raised cosine (RC) filters with various roll-off values (beta). (b) Magnification around a few symbols. The raised cosine filter is normalized for a maximum filter coefficient of 1.

reflections in the channel, symbols will not interfere with each other at the receiver either. But a rectangular pulse requires a large transmit bandwidth.

Proper baseband pulse shaping is thus required to band-limit the transmit spectrum to fit the available channel, minimize out-of-band interference, and improve transmit spectrum efficiency. Besides band-limiting the transmit spectrum, pulse shaping plays a key role in signal detection in digital communications: by combining the transmit pulse-shape filter with a matched filter at the receiver, we can average down channel noise and dramatically improve signal-to-noise ratio (SNR) in the receiver. This technique is known as matched filtering. For zero inter-symbol interference (ISI), a matched filter must satisfy the Nyquist ISI criterion, that is, its impulse response must be zero at all sampling instants except its own where it peaks [4].

A Nyquist pulse-shape filter commonly used in digital communication systems is the raised cosine filter [5]. Assuming a flat channel response and white noise, the response of a Nyquist pulse-shaping function can be split equally between the transmitter and receiver [6]. The raised cosine matched filter is typically implemented by using a square root raised cosine (SRRC) filter at the transmitter and an SRRC filter at the receiver. Although each SRRC filter separately does not satisfy the Nyquist ISI criterion, the combination of two SRRC filters does. Later in this chapter we will have more to say about matched filtering as applied to maximum likelihood RFID reader receivers.

Pulse shaping and matched filtering are often combined with interpolation and decimation, respectively. We accomplish that here by interpolating the baseband data using pulse-shape FIR filters. Figure 5.2 shows a reader PIE sequence containing a message header followed by arbitrary data interpolated with a rectangular pulse filter and raised cosine filters with variable roll-off factors. In this example, the PIE symbols '0' and '1' are coded with two and four bits, respectively, and then interpolated with an up-sampling factor equal to 8, resulting in a duration of 16 samples (Tari) for the PIE symbol '0' and 32 samples (2Tari) for the PIE symbol '1'.

5.2.3   Double-Sideband Amplitude Modulation

Double-sideband amplitude-shift keying (DSB-ASK) modulation is an implementation of amplitude modulation (AM) where the baseband-to-RF conversion produces two frequency-shifted copies of the baseband message signal on either side of the desired RF carrier. In DSB suppressed carrier modulation, the modulated RF signal can be represented in the discrete domain as

$$s[n] = m[n] \cos(2\pi f_c n / F_s), \tag{5.1}$$

where $m[n]$ is the baseband message signal sampled at discrete instants n at a sampling rate equal to F_s, and f_c is the RF carrier frequency. In the digital DSB-ASK defined in ISO 18000-63, the (unfiltered) message signal $m[n]$ can have two levels (see the PIE-encoded data in Figure 5.2). The Fourier representation of $s[n]$ is given by

$$S(f) = \frac{1}{2}[M(f - f_c) + M(f + f_c)], \tag{5.2}$$

where $M(f)$ is the discrete-time Fourier transform of $m[n]$. In DSB modulation, a baseband bandwidth BW yields an RF bandpass bandwidth of $2BW$. For increased spectrum efficiency, the ISO 18000-63 standard specifies two additional modulation formats (see next sections).

5.2.4   Single-Sideband Amplitude Modulation

The ISO 18000-63 standard specifies SSB-ASK modulation, in which the lower or upper sideband of the transmit spectrum is eliminated to improve spectral efficiency. Two approaches can be used to eliminate one of the sidebands, one based on explicit bandpass filtering of the signal, and the other based on signal phasing. Selectively filtering the upper or lower sideband is challenging because of the stringent filter requirements to precisely remove the unwanted sideband while preserving the sideband of interest.

In the second method (signal phasing), the message signal is shifted to the carrier frequency (f_o) without creating the pair of frequency components that are usually

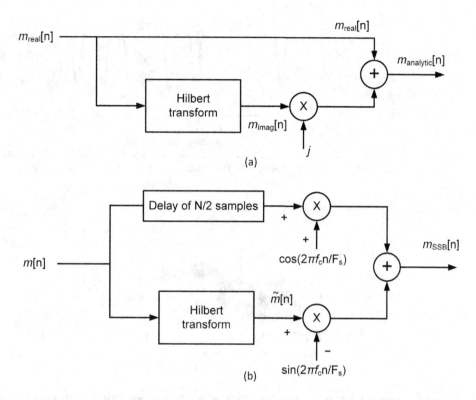

Figure 5.3 (a) Diagram of Hilbert transform. (b) Practical Hilbert transform implementation for generating frequency-shifted SSB modulation. The delay of $N/2$ samples in the upper branch is used to compensate for the delay introduced by the Nth-order FIR Hilbert filter.

present in DSB modulation, avoiding the need for explicit filtering of the lower or upper sideband. This can be achieved by using a Hilbert transform. The ideal Hilbert transform phase-advances the negative frequency components by 90 degrees while phase-delaying the positive frequency components by the same amount [7, 8]. By applying an additional phase shift of 90 degrees $(+j)$ to the output of the Hilbert transformer, the negative frequencies are advanced by 180 degrees relative to the original signal while the initial 90 degree phase-delay in the positive frequencies is reversed. Adding the resulting signal to the original signal (Figure 5.3(a)) cancels out the negative frequency components while adding up the positive frequency components yielding the complex signal (5.3), known as the analytic signal,

$$m_{\text{analytic}}[n] = m_{\text{real}}[n] + jm_{\text{imag}}[n].\tag{5.3}$$

The analytic signal has the following key features: (i) it only exists in the positive Nyquist interval, that is, it has no negative frequency components; (ii) it has half the bandwidth of the original signal; and (iii) it presents twice the amplitude of the original signal.

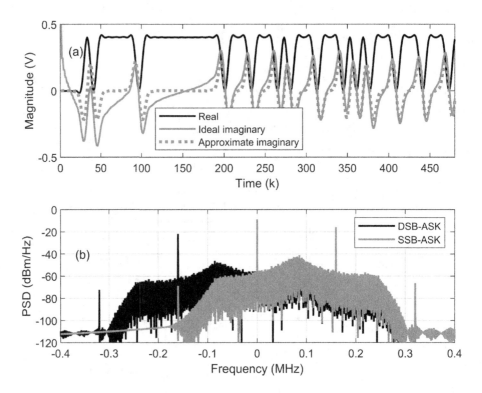

Figure 5.4 PIE-encoded data modulated with DSB-ASK and SSB-ASK. The sequence starts with an ISO 80006-63 PIE header followed by arbitrary data representing a reader command. (a) Time-domain waveforms. (b) Spectral power densities.

A common application of this result to achieve sideband spectrum suppression and frequency shifting is shown in Figure 5.3(b), where the real part of the analytic signal is multiplied by a cosine signal and the imaginary part is multiplied by a sine signal at the desired shifting frequency. In practice, this can be done by using an in-phase and quadrature (IQ) modulator [7]. Typically, a DSB-ASK digital baseband signal is first filtered with a Hilbert filter to create a complex IQ (SSB) signal with suppressed negative frequencies. The resulting SSB-ASK signal is frequency offset to center its spectrum in the middle of the desired channel and then converted to the analog domain to drive an IQ modulator [3].

The MATLAB function *hilbert* can be used to compute the exact analytic signal (5.3), but an ideal Hilbert transform implementation is impractical. The imaginary part of the analytic signal can be approximated using an FIR Hilbert transform filter. Figure 5.4(a) shows the complex waveform of a PIE-encoded SSB-modulated signal, and Figure 5.4(b) compares the spectrum of this signal to that of the same PIE sequence modulated with DSB-ASK. Notice the smaller bandwidth of the SSB-ASK-modulated signal compared to the DSB-ASK-modulated for the same data rate. In both cases, we used a pulse-shaping raised cosine filter with a 0.9 roll-off factor.

5.2.5   Phase-Reversal Amplitude Modulation

To reduce the bandwidth of the transmit signal, the ISO 18000-63 PR-ASK modulation forces a phase inversion in the RF carrier at the beginning of every other baseband symbol (see previous chapter). This leads to an average bandwidth reduction of 50% for the same bit rate compared to the DSB-ASK modulation. To implement the PR-ASK phase inversion, an antipodal baseband signal is required.

Since PR-ASK conveys information in amplitude, PR-ASK-modulated signals produce similar outputs in envelope detectors to standard ASK signals. Therefore, conventional transponder envelope detectors can process PR-ASK-modulated signals without requiring any modification. Figure 5.5 presents the time-domain waveform and spectrum of a PIE-encoded data sequence. As can be seen, a smaller bandwidth can be achieved with PR-ASK modulation compared to DSB-ASK modulation for the same data rate. Here, we used the same filter parameters as in the previous example, but these parameters, which are entered as input to the MATLAB Code 5.1, can be adjusted to meet specific spectrum/mask requirements (see next section).

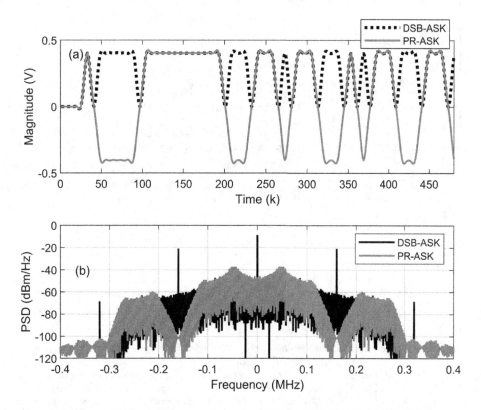

Figure 5.5 PIE-encoded data sequence modulated with DSB-ASK and PR-ASK. (a) Time-domain waveforms. (b) Spectral power densities. The side lobes in the PR-ASK spectrum can be minimized by adjusting transmit parameters, principally the pulse-shaping filter parameters.

Code 5.1 Encoding, modulation, interpolation, and pulse-shape filtering of ISO 18000-63 transmit reader frames (continues on the next page).

```
function [ s_tx] =
EncodeModulateInterpolate(data,mod_type,command_type,beta,tari,tx_filter,span,
shape,norm)
% This function encodes, modulates and interpolates reader command data
% The data are encoded using pulse interval encoding (PIE) and modulated
% using ASK, SSB-ASK or PR-ASK. Interpolation and pulse-shaping are
% performed in a signle step using a rectangular or raised cosine filter
% Inputs:
%    data - binary payload data of the reader command to be transmitted
%    mod_type - modulation type (0: DSB-ASK, 1: PR-ASK, 2: SSB-ASK)
%    command_type - command type (1: Query, 0: all other commands)
%    beta - filter roll-off (raised cosine only)
%    tari - duration of PIE symbol '0'
%    tx_filter - defines transmit filter (0: rectangular pulse, 1: raised cosine)
%    span - filter span in symbols (raised cosine only)
%    shape - filter shape (raised cosine only)
%    norm - nomalize filter gain (raised cosine only)
% Output:
%    s_tx - reader transmit baseband waveform
% Author: Alirio Boaventura

if(mod_type == 2 && tx_filter == 0)
  error('SSB modulation not allowed with rectangular pulse shape');
end

d_0 = tari;              % Duration of PIE symbol '0'
d_pw = tari/2;           % Duration of PIE pulse (PW)
d_1 = 2*tari;            % Duration of PIE symbol '1'
d_av = (d_0+d_1)/2;      % Average transmit symbol duration
tx_dr = 1/d_av;          % Average transmit data rate
Fs = 2.56e6;             % Sampling rate (2.56 Megasample/s)
Ts = 1/Fs;               % Sampling period
n_0 = round(d_0/Ts);     % Number of samples of symbol '0'
n_1 = round(d_1/Ts);     % Number of samples of symbol '1'
n_pw = round(d_pw/Ts);   % Number of samples of PIE pulse
upfac = n_pw;            % Upsampling factor
s_re = 0;                % Real part of the signal
s_im = 0;                % Imaginary part of the signal

% PIE encoding
% Symbol '0' and '1' contain 2 and 3 pulses of duration tari/2,
% respectively. Each symbol will be upsampled by a factor of upfac = 8
s0 = [ 1 0];             % Symbol '0' (duration of tari)
s1 = [ 1 1 1 0];         % Symbol '1' (duration of 2*Tari)

% Define Preamble and FrameSync
d_delim = 12.5e-6;   % Duration of delimiter symbol
d_RTcal = d_0+d_1;   % Duration of reader-to-tag calibration symbol (d_0+d_1)
d_TRcal = 37.5e-6;   % Duration of tag-to-reader calibration symbol
                     % The d_TRcal and the DR parameter in the query command
                     % define the transponder's BLF (see ISO 18000-63, table 6.9)
                     % (e.g., DR = 64/8 and d_TRcal = 33.3e-6 -> max BLF = 640kHz)
```

Code 5.1 *(cont.)*

```
n_delim = round(d_delim/Ts);   % Number of (low) samples of the delimiter
n_RTcal = round(d_RTcal/Ts);   % Number of (high) samples of RTcal
n_TRcal = round(d_TRcal/Ts);   % Number of (high) samples of TRcal
preamb_dsb_ask = [ zeros(1,n_delim/n_pw), s0, ones(1,n_RTcal/n_pw), ...
  zeros(1,n_pw/n_pw), ones(1,n_TRcal/n_pw), zeros(1,n_pw/n_pw)]; % Preamble DSB-
ASK
preamb_pr_ask = [ zeros(1,n_delim/n_pw), s0, -ones(1,n_RTcal/n_pw), ...
  zeros(1,n_pw/n_pw), ones(1,n_TRcal/n_pw), zeros(1,n_pw/n_pw)]; % Preamble PR-
ASK
% DSB-ASK FrameSync
frameSync_dsb_ask = [ zeros(1,n_delim/n_pw), s0, ones(1,n_RTcal/n_pw),
zeros(1,n_pw/n_pw)];
% PR-ASK Frame-sync
frameSync_pr_ask = [ zeros(1,n_delim/n_pw), s0, -ones(1,n_RTcal/n_pw),
zeros(1,n_pw/n_pw)];

% Symbol mapping/Modulation (DSB-ASK, SSB-ASK or PR-ASK)
invert_next = 1;
if(mod_type == 0 || mod_type == 2) % DSB-ASK or SSB-ASK
  if(command_type == 1)            % Start with Preamble
    s_re = [ preamb_dsb_ask];
  else   % Start with FrameSync
    s_re = [ frameSync_dsb_ask];
  end
  for n = 1:length(data)
    if (data(n) == 1)              % Add symbol '1'
      s_re = [ s_re,s1];
    else                           % Add symbol '0'
      s_re = [ s_re,s0];
    end
  end
elseif(mod_type == 1)              % PR-ASK
  if (command_type == 1)
    s_re = [ preamb_pr_ask];       % Start with Preamble
    invert_next = 1;
  else % Start with FrameSync
    s_re = [ frameSync_pr_ask];
    invert_next = 0;
  end
for n = 1:length(data)
    if (data(n) == 1)              % Add symbol '1'
      if(invert_next == 1)
        s_re = [ s_re,-s1];
        invert_next = 0;
      else
        s_re = [ s_re,s1];
        invert_next = 1;
      end
    else                           % Add symbol '0'
      if (invert_next == 1)        % Invert next symbol
        s_re = [ s_re,-s0];
        invert_next = 0;
      else                         % No symbol inversion
        s_re = [ s_re,s0];
```

Code 5.1 (*cont.*)

```
            invert_next = 1;
        end
      end
    end
end
% Interpolation (up-sampling and pulse-shape filtering)
if(tx_filter == 0)         % Rectangular pulse shape
  s_re = rectpulse(s_re,upfac);
elseif(tx_filter == 1)     % Raised cosine filter
  txfilter = comm.RaisedCosineTransmitFilter(...
      'OutputSamplesPerSymbol',upfac,...
      'Shape',shape,'RolloffFactor',beta,...
      'FilterSpanInSymbols',span);
  if(norm == 1)
    % Normalize filter to obtain maximum filter tap value of 1
    coef = coeffs(txfilter);
    txfilter.Gain = 1/max(coef.Numerator);
  end
  s_re = txfilter(s_re.');
  delay = mean(grpdelay(txfilter)); % Compensate for filter delay
  s_re(1:delay) = [];
end
if (mod_type == 2) % Create complex SSB-ASK signal using Hilbert transform
  N = 40;           % Filter order
  delay = N/2;      % Filter delay
  trans_band = 400e3;      % Transition bandwidth
  Hilbert = designfilt('hilbertfir','FilterOrder',N, ...
    'TransitionWidth',trans_band,'SampleRate',Fs,'DesignMethod''equiripple');
  s_im = filter(Hilbert,s_re);
  % compensate for the filter delay
  s_tx = complex(s_re(1:end-delay),s_im(delay+1:end));
else  % Generate complex DSB-ASK or PR-ASK signal
  s_tx = complex(s_re,s_im);
end
end
```

5.2.6   Transmit Spectrum Mask

Unlicensed UHF RFID readers typically employ frequency-hopping spread-spectrum (FHSS) schemes where the transmit frequency is randomly switched among several channels in a designated band. In the US, the FCC defines channels of 500 kHz in the 902–928 MHz band. In Europe, ETSI has traditionally specified channels of 200 kHz in the band 865–868 MHz and has recently introduced additional channels of 400 kHz in an upper 915–921 MHz band. Readers certified for operation in these regions should confine their transmissions to the assigned channels and shall meet local regulations for spread-spectrum channelization [1, 2].

Out-of-channel emissions can disrupt the operation of other RFID readers in adjacent channels and corrupt signals from transponders that are located even tens of meters away [9]. Out-of-band emissions can also interfere with other wireless systems, including users of licensed spectrum. Therefore, RFID readers are required

Table 5.1 Adjacent channel power ratio performance.

Signal \\ Parameters	PIE-encoded, DSB-ASK-modulated	PIE-encoded, DSB-ASK-modulated	PIE-encoded, PR-ASK-modulated
Signal bandwidth (kHz)	160	160	160
Filter	Rectangular pulse	Raised cosine	Raised cosine
Filter roll-off	—	0.53	0.20
Filter truncation (number of symbols)	—	8	16
Channel bandwidth (kHz)	500	500	250
ACPR 1 (dBc)	18.5	56.1	33.6
ACPR 2 (dBc)	21.9	65.4	63.5
ACPR 3 (dBc)	27.1	68.6	67.1

Note: MATLAB code for computing ACPR and performing spectrum mask analysis is provided at www.cambridge.org/9781108489713.

to be tested to ensure they meet emission specifications. The ISO 18000-63 standard specifies transmit spectrum masks for multiple- and dense-interrogator environments in terms of adjacent channel power ratios (ACPR). Readers certified for operation according to this standard shall also meet local regulations for out-of-channel and out-of-band emissions (e.g., those imposed by the FCC and ETSI).

Out-of-band and out-of-channel interference pose an important limitation on the type of encoding, modulation, and maximum data rate allowed to an RFID reader as these parameters determine the bandwidth of the resulting signal. DSB-ASK, SSB-ASK, and PR-ASK modulations can require channel bandwidths that are up to four times, three times, and two times wider than their signal bandwidths, respectively [10]. To meet transmit masks while maximizing spectrum efficiency, adequate baseband pulse-shape filtering and efficient modulation schemes like those discussed in the previous sections are usually required.

Here, we analyze the ISO 18000-63 spectrum mask for multiple interrogator environments for the modulation formats discussed previously. Figure 5.6 depicts the power spectrum density (PSD) of a 160 kHz (Tari $= 6.25$ μs) PIE-encoded DSB-ASK-modulated signal interpolated with a rectangular pulse filter as implemented in the function *EncodeModulate*(). Figure 5.6 also shows the dense-interrogator spectrum mask for a 500 kHz channel (dashed lines). Not surprisingly, the unfiltered ACPR performance of this unfiltered signal does not comply with the ISO 18000-63 spectrum mask (see Table 5.1).

To meet the ACPR requirements of the previous baseband signal in a 500 kHz channel, we filtered the signal with a raised cosine pulse-shaping filter having a roll-off of 0.53 truncated to the duration of eight symbols (see Figure 5.7). This filter is implemented in the function provided in Code 5.1. Figure 5.8 presents the spectrum

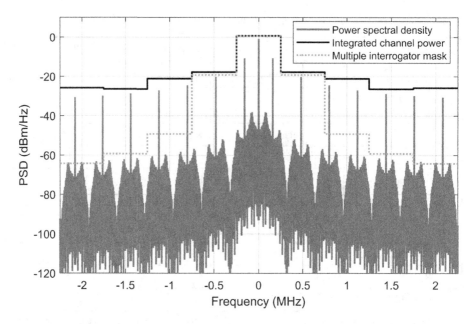

Figure 5.6 PSD of 160 kHz PIE-encoded DSB-ASK-modulated signal with no pulse-shape filtering applied to the signal. The dashed lines correspond to the ISO 18000-63 multiple interrogator spectrum mask and the power is integrated in a 500 kHz channel bandwidth.

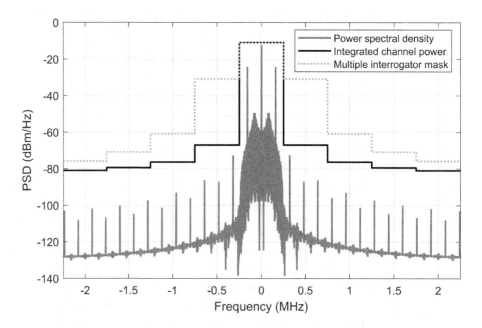

Figure 5.7 PSD of 160 kHz PIE-encoded DSB-ASK-modulated signal filtered using a raised cosine filter with 0.53 roll-off truncated to eight symbols. The dashed lines correspond to the ISO 18000-63 multiple interrogator spectrum mask and the power is integrated in a 500 kHz channel.

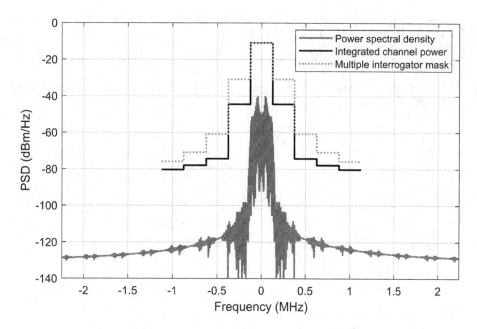

Figure 5.8 PSD of 160 kHz PIE-encoded PR-ASK-modulated signal filtered using a raised cosine filter with 0.2 roll-off truncated to 16 symbols. The dashed lines correspond to the ISO 18000-63 multiple interrogator spectrum mask and the power is integrated in a 250 kHz channel.

mask analysis for a 160 kHz PIE-encoded PR-ASK-modulated signal. By using a raised cosine filter with a roll-off of 0.2 truncated to the duration of 16 symbols, we achieved spectrum mask compliance for a 250 kHz channel. This translates into an improved spectrum efficiency compared to DSB-ASK. Note that:

- The results in Figures 5.6–5.8 do not account for potential non-linearities in the reader transmit power amplifier, which can introduce both in-band and out-of-band distortions. To mitigate spectrum regrowth and other spurious emissions, bandpass filtering should be used at the output of the transmitter.
- Besides meeting spectral mask requirements, transmit waveforms shall also comply with time domain specifications. For example, ISO 18000-63 specifies RF envelope ripple, modulation depth, and so on.
- The examples provided here are for illustration purposes only. For real implementations, the reader should refer to local regulations and other relevant specs. See Chapter 6 for an example of spectrum mask analysis applied to a real RFID reader.

5.3   Transponder Waveform Synthesis

To synthesize transponder waveforms for testing the reader receiver algorithms implemented in the following sections, we implemented the MATLAB function

Code 5.2 Random FM0 data generator

```
function [ bin_data,s_tag,s_cod] =
GenRandFM0Frame(sps,nSymb,snrdB,tx_filter,dc_offset,phase_shift,beta,span,
shape)
% This function generates a BPSK-modulated waveform containing random FM0-encoded
transponder data
% Input:
%   sps - number of samples per FM0 symbol
%   nSymb - number of FM0 symbols
%   snrdB - signal-to-noise ratio in dB (Note: for BPSK, SNR=EbN0=EsN0)
%   tx_filter - transmit filter (0: rectangular 1: raised cosine)
%   dc_offset - complex baseband DC offset
%   phase_shift - constellation phase rotation
%   beta - filter roll-off (raised cosine only)
%   span - filter span in symbols (raised cosine only)
%   shape - filter shape (raised cosine only)
% Outputs:
%   data - random binary transponder data
%   s_tag - transponder FM0 waveform
% Author: Alirio Boaventura

% Generate random binary data
bin_data = randi([ 0 1] ,1,nSymb);
bin_data(1) = 0; % force first bit to zero

% FM0 symbol 0 basis pulse
s0 = [ 1 -1];
% FM0 symbol 1 basis pulse
s1 = [ 1 1];

% Pilot tone
pilot = s0;
for n = 1:11
  pilot = [ pilot,s0];
end

% Preamble pattern
preamb = [ s1 -s0 -s1 s0 -s1 s1];

% FM0 header
s_cod = [ pilot,preamb,-s0];

% FM0 biorthogonal encoding
last_was_pos = 1;
for n = 2:nSymb
  if (last_was_pos == 1)
    if (bin_data(n) == 0)
      s_cod = [ s_cod,-s0];
    else
      s_cod = [ s_cod,-s1];
      last_was_pos = 0;
    end
  else
    if(bin_data(n) == 0)
      s_cod = [ s_cod,s0];
    else
```

Code 5.2 (cont.)

```
      s_cod = [ s_cod,s1] ;
      last_was_pos = 1;
    end
  end
end
% Force a final transition with a symbol '1'
s_cod = [ s_cod -s_cod(end) -s_cod(end)] ;
% Termination sequence to compansate for transmit-receive filer delay
s_cod = [ s_cod, s_cod(end)*ones(1,4*span)] ;

% Interpolation (up-sampling followed by pulse-shape filtering)
% The signal is upsampled at half the desired oversampling because
% each FM0 symbol has been pre-coded with two binary symbols
% (i.e., it is already upsampled by a factor of 2)
if(tx_filter == 1)  % interpolation using raised cosine filter
  % Note that other raised cosine filter design tools could be used to
  % implement this filter (in case the communications toolbox unavailable
  txfilter = comm.RaisedCosineTransmitFilter(...
    'OutputSamplesPerSymbol',sps/2,...
    'RolloffFactor',beta,'Shape',shape,...
    'FilterSpanInSymbols',span');
  s_shaped = txfilter(s_cod.');
else % interpolation using rectangular pulse shape
  s_shaped = [] ;
  for n = 1:length(s_cod)
    s_shaped = [ s_shaped ones(1,sps/2)*s_cod(n)] ;
  end
end
  % Create complex baseband signal
  s_mod = complex(s_shaped,0).';
  % Pass signal through AWGN channel with prescribed SNR
  s_awgn = awgn(s_mod,snrdB,'measured');
  % Apply phase rotation and complex DC offset
  s_tag = s_awgn*exp(j*deg2rad(phase_shift))+dc_offset;
end
```

GenRandFM0Frame() [Code 5.2], which can be used to generate waveforms with random FM0 data and variable parameters. In addition, QPSK-modulated signals can be generated using our custom function *GenRandQPSKFrame*(). This transponder waveform model accounts for channel non-idealities including arbitrary white Gaussian noise (AWGN), constellation phase rotation, and down-conversion baseband DC offset generation. A baseband FM0 waveform with zero DC offset centered about the origin produces a constellation similar to a standard BPSK-modulated signal and can be demodulated in a similar way (transponder signal demodulation is discussed later).

The provided code allows for either rectangular or raised cosine pulse shaping of the generated transponder waveforms. For simplicity, passive-backscatter transponders typically employ on–off modulation and rectangular pulse shaping, but transponder signals are naturally filtered as they go through the communication channel including the reader receiver circuitry, but deliberate transponder baseband pulse shaping has recently been explored to improve the spectrum efficiency of transponder transmissions [11].

Figure 5.9 depicts the constellations of FM0 (BPSK) and QPSK waveforms generated with Code 5.2. Figure 5.9(a) corresponds to on–off rectangular pulse shaping and Figure 5.9(b) corresponds to typical transponder FM0. Figures 5.9(c) and (d) present a 4-QPSK and a 128-QPSK signal, respectively. Although transponders conventionally produce two-state modulation, quadrature backscatter modulation has recently been proposed in the literature. For example, the QPSK signals presented in Figures 5.9(c) and (d) could be generated in practice by using a quadrature backscatter modulator like the one we proposed in Chapter 3. The 128-QPSK signal in Figure 5.9(d) resembles the backscatter chirp spread-spectrum modulation presented in [12].

For the examples in the following sections, we only display the in-phase component of the transponder waveforms but in general the intercepted transponder signal is demodulated using an IQ demodulator and the resulting complex baseband signal contains both in-phase and quadrature components.

5.4   Receiver Signal Processing

In this section, we evaluate the ISO 18000-63 receive protocol using custom MATLAB code. To evaluate the receiver baseband algorithms discussed here, we synthesize FM0 waveforms with variable parameters using the code provided in the previous section. In addition, we tested these algorithms using experimental baseband data acquired from real transponders either with our custom RFID reader discussed in the next section or with standard RF test equipment. These algorithms can be easily extended to Miller encoding and can be reused for custom designs. Applications of some of the approaches discussed here in real RFID readers can be found in the following chapters.

5.4.1   Maximum Likelihood Detection

Maximum likelihood estimation is central to radar and digital communications and can be applied in RFID readers for optimum detection of transponder information. The signal intercepted by an RFID reader receiver contains the transponder message plus AWGN and other distortions introduced by the communication channel, including the reader receiver hardware itself. Thus, the fundamental challenge of the receiver is to recover the transponder message that was transmitted after the signal that carries that message has been corrupted.

Given that the channel-induced noise is a random entity, and the transmitted message contains deterministic symbols, this message can be detected in an optimum manner by evaluating the similarity between the received signal and the transmission symbols [4, 13] (see Figure 5.10). In maximum likelihood estimation, the decision on

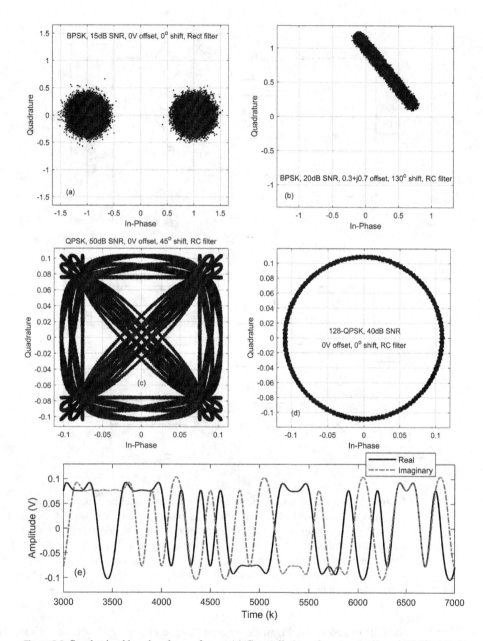

Figure 5.9 Synthesized baseband waveforms. (a) Constellation of centered, aligned BPSK signal. (b) Constellation of BPSK signal containing random DC offset and phase rotation. (c) Constellation of 4-QPSK signal. (d) Constellation of 128-QPSK signal. (e) Time-domain portion of the 4-QPSK signal. The first example uses rectangular pulse shaping (pure on–off modulation) while the other examples apply transmit raised cosine (RC) pulse shaping with 0.5 roll-off.

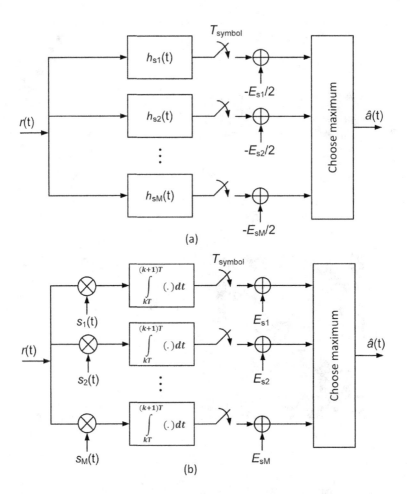

Figure 5.10 (a) Matched filtering realization of maximum likelihood detection for alphabet of *M* transmission symbols. (b) Correlation realization of maximum likelihood detection. The received signal is compared to all the *M* alphabet symbols to determine the most likely transmitted symbol. $r(t)$ and $s_m(t)$ are the received signal and local template symbol corresponding to the *m*th alphabet symbol, respectively, $h_{sm}(t)$ and E_{sm} are the *m*th filter impulse response and symbol energy, respectively, and $\hat{a}(t)$ is the decimated estimate of the transmitted symbol.

the transmitted symbol is made by comparing each received symbol with local copies/ templates of all possible symbols and selecting the one that most closely resembles the received symbol.

The maximum likelihood receiver, also known as minimum Euclidean distance receiver, maximizes the likelihood that a certain symbol was transmitted by minimizing its Euclidean distance to all possible symbol vectors in the alphabet. This requires a priori knowledge of all transmission symbols. Fortunately, the ISO 18000-63 transponder transmission pulses are partially known to the reader receiver (the format of the symbols and their nominal durations are known a priori). In addition, the transponder binary data are encoded using biorthogonal basis waveforms (see Chapter 4). Figure 5.10 shows the structure of typical maximum likelihood receivers based on matched filtering and cross-correlation.

By matching the receive filter to the transmit pulse-shape filter, that is, making its impulse response equal to the time-reversed and time-shifted version of the impulse response of the transmit filter, one can minimize ISI and maximize SNR [4]. The benefits of matched filtering can be more intuitively understood by observing that: (i) each matched filter pulse has a low-pass frequency response that filters out high-frequency noise; (ii) the matched filter integrates the signal over the symbol period, which effectively averages down zero-mean AWGN; (iii) as a result of (i) and (ii), the subsequent blocks in the receiver chain experience a stronger signal at the optimum sampling instant compared to directly sampling the original signal; and (iv) the use of orthogonal pulses minimizes the likelihood of false detection.

Another way of evaluating the similarity between two signals is through cross-correlation [4]. As opposed to the matched filtering implementation (Figure 5.10(a)), which uses linear filtering, the cross-correlation method uses non-linear operations, namely multiplication followed by integration (see Figure 5.10(b)). This method is simpler to implement and produces the same output as matched filtering at the optimum sampling points but may differ elsewhere depending on the matched filtering pulse format. Throughout this book, we will refer to correlation and matched filtering inter-changeably, but the reader should keep in mind that they are not necessarily the same.

Figure 5.11 presents eye diagrams of a rectangular filter and a raised cosine matched filter consisting of a transmit and receive SRRC filter. Notice the greater opening in the eye diagrams at the matched filter output and much better-defined sampling instants compared to the original signal. This translates into improved SNR delivered to the subsequent blocks in the receiver chain.

So far, we have assumed that the sampling clock is perfectly synchronized with the received signal. But the signal intercepted by the reader usually presents frequency and phase deviations due to imperfections like clock jitter, phase noise, and channel dispersion. To maximize SNR at the matched filter's output and minimize bit error rate (BER), it is crucial that each symbol be sampled at the optimum sampling instant; sub-optimal sampling can lead to significant degradation in SNR and defeat the purpose of using a matched filter. This is the motivation for the inclusion of a timing recovery feedback loop that tracks and corrects frequency and phase deviations in the received signal. In the following sections, we present a comprehensive discussion of timing recovery suitable for RFID applications.

5.4.2   Symbol Synchronization

In digital communications systems, proper synchronization between the transmitter and receiver is key to retrieving the transmitted message. Conventional wireless systems typically employ several levels of synchronization, including at the symbol, carrier, and frame level. Some systems may use dedicated synchronization channels in addition to the main data transmission channel, but this approach leads to increased

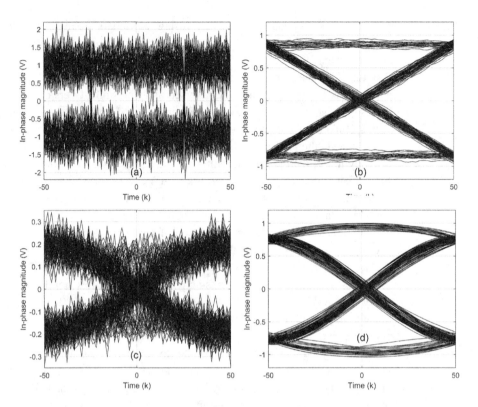

Figure 5.11 Eye diagram analysis of FM0 waveforms oversampled at 100 samples per symbol in an AWGN channel with 6 dB SNR using a rectangular and SRRC matched filter with a duration of half the FM0 symbol period. (a) Rectangular matched filter input. (b) Rectangular matched filter output. (c) SRRC matched filter input. (d) SRRC matched filter output for a filter roll-off of 0.7.

system complexity and poor energy and spectrum efficiency. In many modern systems, including RFID systems, the synchronization information is transmitted along with the data signal.

Carrier synchronization is intended to frequency- and phase-align the receiver to the transmitter RF carrier [14]. This level of synchronization is usually not required in backscatter RFID readers based on direct conversion, as the transmitter and receiver share the same local oscillator. But frequency deviations between transmitted and backscattered signals can still occur in applications like vehicular identification due to doppler shift [15]. These deviations are usually small and do not significantly affect the receiver operation. Frame synchronization, which aims to estimate the starting position of a frame in the received signal, is discussed later in this chapter.

A key challenge in recovering the transponder message in the receiver is to find the optimal instant to sample the received signal to maximize the channel SNR and minimize the ISI and BER. This can be accomplished by using symbol synchronization, also referred to as symbol recovery, timing recovery, or clock recovery.

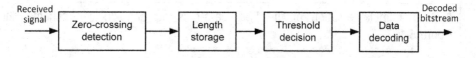

Figure 5.12 FM0 symbol demodulation based on a simple zero-crossing detection scheme.

A simple approach to estimating the timing of FM0- or Miller-encoded signals consists of using a zero-crossing detection scheme like the one shown in Figure 5.12 [16]. In this approach, the transponder data are decoded by detecting transitions (zero-crossings) in the incoming baseband signal and comparing the time elapsed between these transitions to a predefined threshold (see Code 5.3). This approach presents sub-optimal performance as it is very sensitive to amplitude noise (see Section 5.5), but it is simple and requires minimal computation and no explicit symbol synchronization.

For optimum performance in systems with self-timing (i.e., with no dedicated channel for synchronization), the receiver must be able to accurately estimate the correct symbol rate and sampling instant within each symbol period and to dynamically update these parameters to compensate for frequency and phase fluctuations during transmissions. This process typically involves estimating the clock phase error and correcting the clock phase based on the estimated error. In the following sections, we discuss optimum clock recovery techniques based on phase-locked loops (PLLs).

5.4.3 Phase-Locked Loop Clock Recovery

Phase-locked loops are used in a variety of applications in wireless communications systems, including frequency synthesis and carrier and clock synchronization. Here, we discuss the digital PLL (DPLL) technique for timing and symbol recovery in RFID reader receivers. DPLL symbol synchronization uses a feedback loop to track and correct timing error in the received signal at the output of a matched filter (see Figure 5.13). As mentioned before, optimizing the sampling instants is key to maximizing channel SNR and improving reception BER performance.

A typical DPLL system (the dashed region in Figure 5.13(a)) comprises three main components: timing error detector (TED), interpolator (and interpolation controller), and loop filter. Based on some criterion that may depend on present and past symbol decisions (in the case of decision-directed schemes), the TED block estimates an unknown error offset $e(k)$ and applies a scaled version of this error to compute a timing error offset $\hat{\tau}_k$ that is used to update the interpolation instant. The offset error estimate can be obtained using different algorithms [17–20].

Since the incoming signal is discretized and the symbol rate is typically not synchronized with the ADC sampling rate, the desired optimum sampling instant is generally not aligned with an available sample. The role of the interpolator is to resample the output of the matched filter at an arbitrary instant $t = kT_{\text{symb}} + \hat{\tau}_k$ based on the available samples (see Figure 5.14). In this sense, the interpolator effectively causes a fractional delay to the symbol sampling. The desired sample $r\left(kT_{\text{symb}} + \hat{\tau}_k\right)$

Code 5.3 Implementation of the FM0 zero-crossing demodulation scheme shown in Figure 5.12

```
function [ data] = ZCrossFM0Demod(s_rx,sps,nSymbs)
%  This function demodulates an FM0 signal using zero crossing detection.
%  The signal timing is first recovered by counting the number of samples between
%  consecutive signal transitions and the data is then decoded by evaluating the
%  stored lengths using a pre-defined threshold.
%
%  Input:
%    s_rx - received FM0 signal
%    sps - number of samples per symbol
%    nSymb - expected number of symbols
%  Output:
%    data - decoded binary data
%  Author: Alirio Boaventura

  N = length(s_rx);
  data = zeros(1,nSymbs);        % Initialize output data vector
  tol = 0.2;                     % Allowed phase deviation relative to nominal
sps (20%)
  last_was_pos = 0;              % Sign of last processed sample
  nCrosses = 0;                  % Number of zero crossings
  nSamps = 0;                    % Number of samples processed since last zero
crossing
  MaxIters = 100;                % Max number of iterations searching for
preamble
  d_v_max = 1.5*sps*(1 + tol);   % Max duration of FM0 symbol 'V'
  d_v_min = 1.5*sps*(1 - tol);   % Min duration of FM0 symbol 'V'
  d_1_max = sps*(1 + tol);       % Max duration of FM0 symbol '1'
  d_1_min = sps*(1 - tol);       % Min duration of FM0 symbol '1'
  d_half0_max = sps*(1 + tol)/2; % Max duration of half FM0 symbol '0'
  d_half0_min = sps*(1 - tol)/2; % Min duration of half FM0 symbol '0'
  clk = [];
  %% Zero-crossing detection and length storage (timing recovery)
  for n = 1:N
    nSamps = nSamps+1;
    % Negative and positive going zero crossings are detected when x(n)*x(n+1) < 0
    if((last_was_pos && ~IsPositive(s_rx(n))) || (~last_was_pos && IsPositive
(s_rx(n))))
      nCrosses = nCrosses+1;    % Increment number of zero crossings detected
      clk(nCrosses) = nSamps;   % Save distance between consecutive zero crossings
      nSamps = 0;               % Reset number of samples since last zero crossing
    end
    last_was_pos = IsPositive(s_rx(n));
  end

  %% Preamble detection
  M = length(clk);
  n = 1;
  while(n < MaxIters)
    if(n+6 > M)
      break
    end
    P1 = ((clk(n) > d_1_min) && (clk(n) < d_1_max));            % Symbol '1'
    P2 = ((clk(n+1) > d_half0_min) && (clk(n+1) < d_half0_max)); % Half symbol '0'
```

Code 5.3 (*cont.*)

```
      P3 = ((clk(n+2) > d_half0_min) && (clk(n+2) < d_half0_max));  % Half symbol '0'
      P4 = ((clk(n+3) > d_1_min) && (clk(n+3) < d_1_max));          % Symbol '1'
      P5 = ((clk(n+4) > d_half0_min) && (clk(n+4) < d_half0_max));  % Half symbol '0'
      P6 = ((clk(n+5) > d_v_min) && (clk(n+5) < d_v_max));          % Symbol 'V'
      P7 = ((clk(n+6) > d_1_min) && (clk(n+6) < d_1_max));          % Symbol '1'
      if(P1 && P2 && P3 && P4 && P5 && P6 && P7) % Check for occurrence of preamble
pattern
          n = n+7; % Starting position of payload passed to the following loop
          break;
      end
      n = n+1;
  end
  %% Threshold decision and symbol demapping
  for m = 1:nSymbs
    if(n > M)
      break;
    end
    % Symbol '1' is detected if (1-tol)*sps < clk(n) < (1+tol)*sps
    if((clk(n) > sps*(1-tol)) && (clk(n) < sps*(1+tol)))
        data(m) = 1;
        n = n+1;
    % Symbol '0' is detected otherwise
    else
        data(m) = 0;
        n = n+2;   % Increment twice as symbol '0' has two transitions
    end
  end
end
```

is then passed to the decision block. Linear interpolation is used here for simplicity, but for improved performance more advanced approaches (e.g., a piecewise parabolic interpolator filter [21]) should be considered.

The control signal for the interpolator is obtained by filtering the error signal $e(k)$ with a proportional integrator (PI) loop filter having a proportional gain G_p and integral gain G_i. The PI loop filter tracks both the phase and accumulated phase errors, and governs the dynamics of the feedback loop, determining the characteristics of the correction including locking time, responsiveness, and damping [4]. Proper selection of the filter parameters is key for achieving optimum performance. For example, excessive overshooting or under-shooting of the desired correction can lead to sub-optimal performance. When the correction is implemented properly, the timing error is effectively driven to zero in a timely fashion, causing the DPLL to lock and the received eye diagram to open.

The interpolation controller controls the interpolation instants by providing the interpolator with the basepoint, which corresponds to the nearest sample to the interpolant, and the fractional interval (see Figure 5.13(b)). In Code 5.4, we present a simplified implementation of DPLL symbol recovery based on the Mueller and Muller timing estimation method. For optimal loop filter and interpolation control design, see [4, 14, 21].

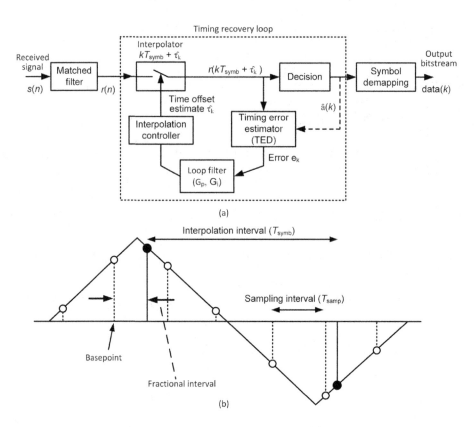

Figure 5.13 (a) DPLL timing recovery. (b) Illustration of interpolation. The interpolation process involves two distinct sampling spaces, one at instant n at the oversampling rate $(1/T_{\text{symb}})$ and the other at instant k at the symbol rate $(1/T_{\text{samp}})$. In general, these frequencies are not synchronous and do not need to be commensurate (i.e., the sampling rate does not need to be a multiple of the symbol rate).

5.4.4   Timing Error Estimation

The timing error $e(k)$ can be estimated using different techniques, including early–late gate [14, 17], zero-crossing detection [4, 16], Gardner [18], and Mueller and Muller [19, 20]. Depending on whether knowledge of the transmitted signal or channel estimation is used to produce the timing error estimate, these techniques are classified as decision-directed (zero-crossing and Mueller and Muller) or non-data-aided (Gardner and early–late gate). Here, we apply the Mueller and Muller method [21], which requires only one sample per symbol and knowledge of the previous symbol to estimate the timing error (see Figure 5.13). The timing error estimate for a complex signal according to the Mueller and Muller method [21] is

$$
\begin{aligned}
e(k) = \ &\widehat{a}_I(k-1)r_I\big(kT_{\text{symb}} + \widehat{\tau}_k\big) - \widehat{a}_I(k)r_I\big((k-1)T_{\text{symb}} + \widehat{\tau}_k\big) \\
&+ \widehat{a}_Q(k-1)r_Q\big(kT_{\text{symb}} + \widehat{\tau}_k\big) - \widehat{a}_Q(k)r_Q\big((k-1)T_{\text{symb}} + \widehat{\tau}_k\big),
\end{aligned}
\tag{5.4}
$$

Code 5.4 DPLL clock recovery loop based on Mueller and Muller error estimation.

```
function [ s_clk, s_match] =
DPLLMnMClockRecovery(s_rx, sps, Gp, Gi, rx_filter, span, shape, beta)
%  This function recovers the clock/timing of an FM0 signal using
%  a digital phase-locked loop (DPLL). The timing error is estimated
%  using the Mueller & Muller method. For details on this method refer to:
%  K. Mueller and M. Muller, "Timing Recovery in Digital Synchronous Data Receivers,"
%  in IEEE Transactions on Communications, vol. 24, no. 5, pp. 516-531, May 1976.
%  and G. Danesfahani and T. Jeans, "Optimisation of modified Mueller and Muller
algorithm",
%  in Electron. Lett., vol. 31, no. 13, pp. 1032-1033, Jun. 1995.
% Inputs:
%  s_rx - transponder FM0 signal to be demodulated
%  sps - nominal number of samples per FM0 symbol
%  Gp - proportional gain of PLL loop filter
%  Gi - integral gain of PLL loop filter
%  rx_filter - receiver matched filter (0: rectangular pulse, 1: raised cosine)
%  span - filter span in symbols (raised cosine only)
%  shape - filter shape (raised cosine only)
%  beta - filter roll-off (raised cosine only)
% Outputs:
% s_matched - matched filter output
% s_clk - decimated clock signal containing two BPSK pulses per FM0 symbol
% Author: Alirio Boaventura
  s_match = [];      % Initialize matched filter output
  s_clk = [];        % Initialize clock signal
  Tsymb = sps/2;     % Nominal symbol interpolation interval assuming each FM0 symbol
                     % contains two BPSK symbols with duration sps/2
  rPrev = 0;         % Matched filter output at previous interpolation instant r(k-1)
  rCurrent = 0;      % Matched filter output at current interpolation instant r(k)
  aPrev = 0;         % Decision on matched filter output at previous instant k-1
  aCurrent = 0;      % Decision on matched filter output at current interpolation instant k
  cumulError = 0;    % Cumulative error cum_e(k)
  error = 0;         % Timing error e(k)
  tau = 0;           % Timing offset t(k)
  Tinterp = 0;       % Symbol interpolation instant
  % Design receiver matched filter
  % The receiver matched impulse response typically is a time-reversed
  % version of the impulse response of the transmit pulse-shape filter, but
  % since the transmit pulse is symmetric, time flipping is not required
  if (rx_filter == 0)     % Rectangular pulse shape
    b_match = ones(1, Tsymb);
  elseif (rx_filter == 1) % Raised cosine filter
    % Any raised cosine filter design function
    % can be used to implement this filter
    rxfilter = comm.RaisedCosineReceiveFilter(...
    'Shape', shape, 'RolloffFactor', beta,...
    'FilterSpanInSymbols', span,...
    'InputSamplesPerSymbol', sps/2,...
    'DecimationFactor', 1);  % We do not decimate the received signal here
                             % Decimation is performed in the clock recovery loop
    b_match = coeffs(rxfilter).Numerator;
  end

  % Compute matched filter output
  if(rx_filter == 0 || rx_filter == 1)
```

Code 5.4 (*cont.*)

```
    s_match = filter(b_match,1,s_rx);
  else
    s_match = s_rx;
  end
  s_match = s_match./max(s_match);    % normalize
  % Clock recovery loop
  for k = 1:round(length(s_rx)/Tsymb)
    % Interpolation controller
    Tinterp = k*Tsymb+tau;            % Update interpolation sampling time
    tBase = floor(Tinterp);           % Compute interpolation basepoint
    tFrac = Tinterp-tBase;            % Compute interpolation fractional interval

    % Interpolate matched filter output
    if(tBase>0 && tBase<length(s_match)-1) % Boundary check
       rCurrent = Interpolate(s_match(tBase),s_match(tBase+1),tFrac);   % Linear
interpolation
    else
       rCurrent = 0;
    end

    % Apply hard decision to interpolated matched filter output
    s_clk(k) = 2*(rCurrent > 0)-1;

    % Save decimated sample
    aCurrent = s_clk(k);
    % Estimate timing error using Mueller and Muller method
    error = (3*Tsymb/16)*(rCurrent*aPrev-rPrev*aCurrent);

    % Compute timing offset
    tau = (tau+Gp*error+Gi*cumulError)*1e-3;

    % Save current matched filter output and interpolated decision
    rPrev = rCurrent;
    aPrev = aCurrent;

    % Update cumulative error
    cumulError = cumulError+error;
  end
end
```

Where $r_I(kT_{symb}+\widehat{\tau}_k)$ and $r_Q(kT_{symb}+\widehat{\tau}_k)$ are the in-phase and quadrature components of the input signal to the TED, T_{symb} is the nominal symbol rate, $\widehat{\tau}_k$ is the estimated timing error at instant k, and $\widehat{a}_I(k)$ and $\widehat{a}_Q(k)$ are the estimates of $r_I(kT_{symb}+\widehat{\tau}_k)$ and $r_Q(kT_{symb}+\widehat{\tau}_k)$, respectively. Figure 5.14 illustrates the error estimation for different situations: the timing error estimation algorithm returns $e=0$ when no timing adjustment is required for the next symbol, $e>0$ if a timing advance is required, or $e<0$ if a timing delay is required.

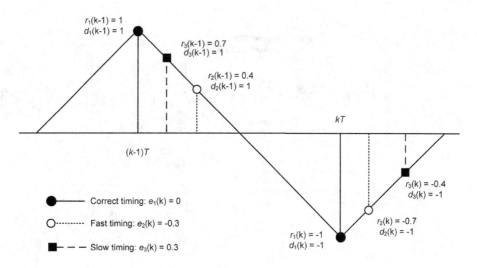

Figure 5.14 Examples of Mueller and Muller error estimation for several cases.

5.4.5   Frame Synchronization

Frame synchronization serves multiple purposes in wireless communications, principally signal detection, channel estimation, and time and frequency offset estimation [22]. Here, we discuss time offset estimation, which is of most interest in typical backscatter RFID systems. The received data frame is generally offset by an unknown delay in the signal intercepted by the receiver (see Figure 5.15(a)). The primary goal of frame synchronization is to estimate this delay and the starting position of the data frame in the received signal. This can be done by prepending a known pattern to the transmitted signal, called the preamble, that can be used in the receiver to estimate the start of the frame.

To improve frame synchronization accuracy, preamble sequences have unique correlation properties. For example, narrowband communication systems commonly employ Baker codes, which have minimal off-peak correlation [4]. In ISO 18000-63, FM0 frame synchronization relies on the use of the preamble sequence '1010V1' leading the transponder payload transmissions, where the symbol 'V' is a code violation that is deliberately introduced to lower the probability of incorrect preamble detection. FM0 preamble detection is illustrated in Figure 5.15.

One way to detect the preamble at the receiver is to perform a cross-correlation between the received waveform and a local copy of the preamble pattern waveform for a variable correlation time lag. In this approach, the starting position of the frame is estimated by finding the time lag that maximizes the cross-correlation between the two signals (see Figure 5.15). Another approach used in wireless communications consists of using a specific sequence of bits to mark the start of a frame and performing frame synchronization after symbol demodulation. But this technique can only be applied if the preamble and payload data use the same modulation format [4].

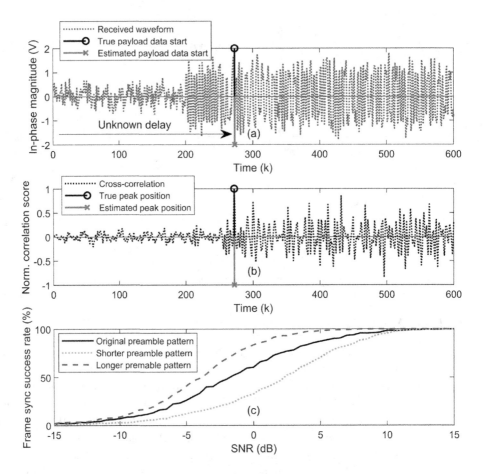

Figure 5.15 Illustration of frame synchronization. (a) Estimated starting position of payload data in the received waveform. (b) Estimated cross-correlation peak. (c) Preamble detection probability for different preamble lengths. Longer preamble patterns improve detection probability and frame synchronization accuracy.

Cross-correlation can be performed either in the sampling space (i.e., in the over-sampled signal) or in the symbol space (i.e., after the received signal has been decimated to the symbol rate). The former approach can be efficiently implemented in software like MATLAB but can become impractical in custom hardware for large oversampling rates, therefore down-sampling of the received signal is often required to reduce the computation burden. In the examples presented in Figure 5.15, we down-sampled the signal to four samples per symbol.

For optimum performance in the calculation of cross-correlations of large sequences, MATLAB's *xcorr*() function uses FFT to exploit the relationship between cross-correlation and convolution and the fact that the latter corresponds to a simple multiplication in the frequency domain [4]. However, since this approach requires the shorter signal to be padded with zeros to match the length of the other signal, the size of the problem increases unnecessarily for sequences with significantly different lengths, as is the case in the preamble detection problem.

A more efficient way of implementing the frame synchronization cross-correlation is to apply a filter to the received signal with an impulse response that is the time-reversed version of the preamble pattern, as in the function *SampleRateFM0FrameSync()* that we provide in Code 5.5. By using a proper oversampling rate, this function can be used for frame synchronization in both the sample and symbol space. For instance, BPSK frame synchronization in the symbol space can be achieved using this function with the number of samples per symbol parameter (sps) set to 1. For the FM0 signal, frame synchronization in the symbol space can be done by using an oversampling of two samples per symbol, assuming that each FM0 symbol is coded with two BPSK pulses (more detail is given later).

In Code 5.6, we present an elegant implementation of FM0 frame synchronization in the symbol space that does not involve cross-correlation and is instead based on an efficient recursive examination of the clock signal.

Besides the preamble, the ISO 18000-63 standard specifies an optional pilot tone with 12 FM0 zero symbols leading the transponder transmissions (see Chapter 4). This pilot tone can be used to implement an ancillary frame synchronization

---

Code 5.5 FM0 frame synchronizer in the sample space.

```
function [ pay_start,s_corr] = SampleRateFM0FrameSync(s_rx,sps)
% This function locates the FM0 preamble pattern in the received signal in the
% sampling rate domain and returns the start position of the payload data
%  Inputs:
%     s_rx - received signal
%     sps - number of samples per symbol
%  Outputs:
%     pay_start - starting position of payload data
%     s_corr - cross-correlation score
%  Author: Alirio Boaventura

  s1 = ones(1,sps);                          % FM0 symbol '1'
  if(mod(sps,2)== 0)                         % Even sps
    s0 = [ ones(1, sps/2), -ones(1, sps/2)]; % FM0 symbol '0'
  else                % Odd sps
    s0 = [ ones(1, floor(sps/2)), ...
        0, -ones(1, floor(sps/2))];          % FM0 symbol '0'
  end
  fm0_preamb = [ s1 -s0 -s1 s0 -s1 s1];      % FM0 preamble pattern

  % Perform cross-correlation efficiently by using filter whose impulse
  % response corresponds to the time-reversed preamble pattern
  s_corr = conv(s_rx,flip(fm0_preamb));
  s_corr = s_corr./max(s_corr);   % Normalize
  pay_start = (find(s_corr == max(s_corr))) + 1; % Return starting position of
payload data
end
```

Code 5.6 Efficient recursive FM0 frame synchronizer at the symbol level.

```
function [ payload_start] = SymbolRateFM0FrameSync(s_clk,stop,k)
%   This function searches for the FM0 preamble pattern in the clock signal using
an efficient
%   recursive approach and returns the start position of the payload data in the
received signal
%   Inputs:
%      s_clk - previously recovered clock signal
%      stop - recursive control variable
%      k - sample index
%   Outputs:
%      payload_start - start of payload data
%   Author: Alirio Boaventura

  MaxIters = 100; % Maximum iterations searching for preamble
  if nargin == 1, k = 1; stop = 0; end
  if ~stop % Search for preamble symbols
      P(1) = (s_clk(k)>0 && s_clk(k+1)>0) || (s_clk(k)<0 && s_clk(k+1)<0);
% Symbol '1'
      P(2) = (s_clk(k+2)>0 && s_clk(k+3)<0) || (s_clk(k+2)<0 && s_clk(k+3)>0);
% Symbol '0'
      P(3) = (s_clk(k+4)>0 && s_clk(k+5)>0) || (s_clk(k+4)<0 && s_clk(k+5)<0);
% Symbol '1'
      P(4) = (s_clk(k+6)>0 && s_clk(k+7)<0) || (s_clk(k+6)<0 && s_clk(k+7)>0);
% Symbol '0'
      P(5) = (s_clk(k+8)>0 && s_clk(k+9)>0) || (s_clk(k+8)<0 && s_clk(k+9)<0);
% Symbol 'V'
     P(6) = (s_clk(k+10)>0 && s_clk(k+11)>0) || (s_clk(k+10)<0 && s_clk(k+11)<0);
% Symbol '1'
      stop = ((sum(P) == length(P)) || (k > MaxIters) || (k+11 > length(s_clk)));
      payload_start = SymbolRateFM0FrameSync(s_clk,stop,k+1);
  else   % Return start of payload data
      payload_start = k + 11;
  end
end
```

---

mechanism, but it adds communication overhead. At the beginning of each inventory round, the reader can command the transponder to not use a pilot tone in its transmissions.

5.4.6 The Full System

ISO 18000-63 Miller and FM0 use a biorthogonal encoding scheme where each binary symbol is associated with a pair of antipodal basis functions (see the previous chapter and [1, 2]). This leads to a receiver structure like the one in Figure 5.10 having four branches ($M = 4$) where each branch is matched to one of the four transponder symbols and the decision is based on the maximum likelihood criterion discussed previously (an implementation of this technique can be found in [23]).

A simpler implementation of the FM0 signal demodulation, which is used in this chapter, is to treat the incoming signal as a BPSK-modulated signal. In this approach, we filter the received signal using a single matched filter with a duration equal to half the FM0 symbol period, then we demodulate the matched filter output using the DPLL clock recovery method discussed previously (Code 5.4), and we demap the resulting clock signal assuming that each FM0 symbol contains two BPSK symbols with half the FM0 symbol duration (Code 5.7). Figure 5.16 presents the waveforms of the DPLL scheme of Figure 5.13 for an AWGN channel and different transmitter–receiver filter configurations, and Figure 5.17 presents typical signal constellations.

The method just described does not perform as well as the maximum likelihood receiver of Figure 5.10 as implemented, for example, in [23], but it is much faster and requires fewer resources. Moreover, this approach does not suffer from the systematic asymmetric error incidence between symbols '0' and '1' previously reported in the literature [24].

In most of the examples presented in this chapter, we have considered only the in-phase component of the transponder baseband signal for simplicity. But as stated earlier, the transponder baseband signal retrieved by the reader IQ demodulator is typically complex, having in-phase and quadrature components. After centering the received signal about the origin and aligning it with the in-phase axis, we obtain a constellation of a BPSK signal (see Figure 5.17). One way to align the constellation is to combine the in-phase and quadrature components, converting the complex signal into a real signal. This approach is followed in the RFID reader design presented in the next chapter, where the complex signal is combined in the analog domain.

Code 5.7 FM0 symbol demapper considering two BPSK pulses per FM0 symbol.

```
function [ data] = FM0SymbolDemap(s_clk,pay_start,nSymbs)
%   This function demaps the data symbols in the recovered BPSK clock signal
%   Inputs:
%     s_clk - previously recovered clock signal
%     pay_start - starting position of payload data
%     nSymbs - expected number of symbols
%   Outputs:
%     data - decoded binary data
%   Author: Alirio Boaventura

  N = length(s_clk);
  m = 1;
  data = zeros(1,nSymbs);
  for n = pay_start:2:pay_start+2*(nSymbs-1)
    if(n >= N)
      break;
    end
    data(m) = ~((IsPositive(s_clk(n)) && ~IsPositive(s_clk(n+1)))...
         || (~IsPositive(s_clk(n)) && IsPositive(s_clk(n+1))));
    m = m+1;
  end
end
```

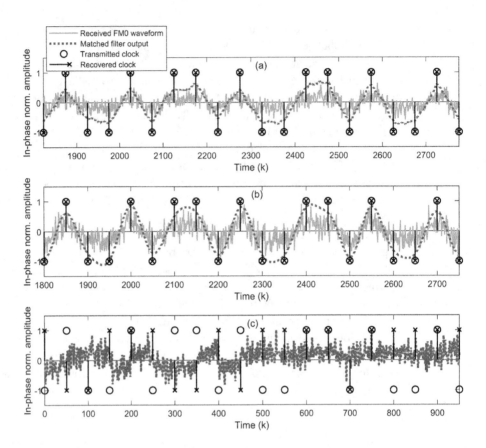

Figure 5.16 FM0 demodulation using the DPLL Mueller and Muller clock recovery implementation presented in Code 5.4. The FM0 frame containing 10 symbols oversampled at 100 samples per symbol is passed through an AWGN channel with a −4 dB SNR. (a) Rectangular matched filter. (b) Square root raised matched filter. (c) No filter used at the receiver. This scenario results in a recovered clock with several symbols in error for the same SNR used in (a) and (b).

5.5   Performance Evaluation

To evaluate the performance of the demodulation methods discussed in this chapter, we synthesize transponder FM0 waveforms with prescribed channel SNR, demodulate them using the zero-crossing detection and DPLL Mueller and Muller algorithms, and compare the obtained BER and transponder reception success rate (RSR). We define the RSR as

$$\text{RSR } (\%) = \frac{\text{Number of successfully demodulated frames}}{\text{Total number of received frames}} \times 100. \qquad (5.5)$$

For baseline comparison, we use BPSK demodulation with ideal sampling and hard decision, and we define the SNR per bit (E_b/N_0) for the oversampled signals as [25]

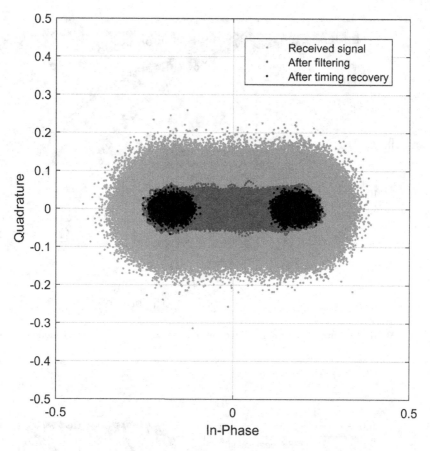

Figure 5.17 Typical constellations involved in the clock recovery scheme illustrated in Figure 5.13. The received signal, containing 1e4 FM0 symbols oversampled at 100 samples per symbol, was passed through an AWGN channel with a 6 dB SNR. In this case, we performed the clock recovery using the Mueller and Muller implementation of MATLAB's communications toolbox.

$$\frac{E_b}{N_0} = \text{SNR} + 10\log_{10}(k) - 10\log_{10}\left(\frac{T_{\text{symb}}}{T_{\text{samp}}}\right), \tag{5.6}$$

where T_{symb} is the BPSK symbol period, T_{samp} is the sampling interval, and k is the number of bits per symbol. The FM0 modulation can be viewed as a BPSK modulation with a code rate of 0.5. Thus, k is set to 1 for the BPSK case and 0.5 for the FM0 case. For the RSR computation, we consider 1e4 frames each containing an FM0 pilot tone and preamble followed by 200 FM0 payload data symbols, which are oversampled at 12 samples per symbol. A frame is successfully received if its preamble and payload data are correctly detected and demodulated. The BER computation is based only on the payload data and is completed when either 200 errors occur or 1e7 bits are received. From the BER and RSR performance results presented in Figures 5.18 and 5.19, the following observations can be made:

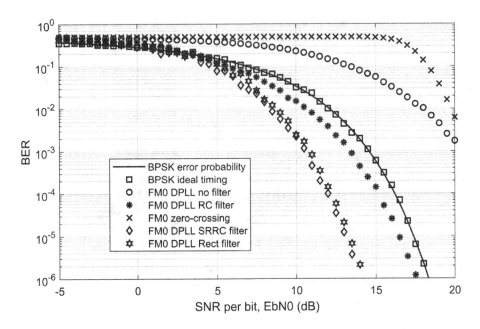

Figure 5.18 Simulated BER as a function of SNR per bit for different demodulation schemes.

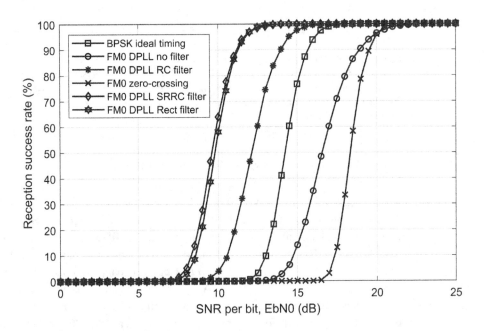

Figure 5.19 Simulated reception success rate as a function of SNR per bit for different demodulation schemes.

- Not surprisingly, Mueller and Muller DPLL demodulation outperforms zero-crossing detection demodulation, given its ability to correct timing error and track optimal sampling instants. Moreover, matched filtering can dramatically improve channel SNR in DPLL receivers, enabling the detection of information signals that are even below the noise floor.
- While zero-crossing detection requires an E_b/N_0 of 18.3 dB (SNR = 7.5 dB) for a 50% reception success rate, the Mueller and Muller DPLL with matched filtering can achieve the same performance at an E_b/N_0 of only 9.6 dB (SNR = -1.2 dB).
- Due to their zero-ISI nature, the rectangular and SRRC matched filters give similar results and largely outperform the other filter topologies. Even though rectangular pulse-shape filters are not practical, they are helpful for benchmarking performance. For simplicity, commercial backscatter transponders typically do not use pulse shaping, but recent work has explored this possibility to improve the spectrum efficiency of transponder transmissions [11].

In low-power and low-complexity applications, zero-crossing detection is preferred over DPLL at the expense of performance. Note that the zero-crossing demodulator used in these experiments used no filter and that low-pass filtering the received signal prior to zero-crossing detection can improve performance. Other symbol synchronization schemes like polyphase filter banks [24, 26] can achieve superior performance compared to the Mueller and Muller method at the expense of increased complexity and computation resources. Here, we focus on FM0 encoding, but other schemes like Miller encoding can outperform FM0 in terms of BER performance (typically, BER improves with the number of Miller subcarrier cycles [27]).

5.6   MATLAB Code

Table 5.2 summarizes the main MATLAB code provided in this chapter for the reader transmit and receive digital signal processing chains, as well as transponder waveform synthesis.

5.7   Exercises

The following exercises relate to transmit spectrum masks and clock recovery previously discussed in this chapter.

5.7.1   Transmit Spectrum Mask

1. Use the provided function *EncodeModulateInterpolate*() with proper input parameters to generate a PIE-encoded SSB-ASK-modulated signal and perform the spectrum mask analysis of this signal for the dense-reader environment requirements specified by the ISO 18000–63 standard as previously done for

Table 5.2 Summary of the main MATLAB code provided in this chapter.

Function	Description
EncodeModulateInterpolate()	Performs reader transmit encoding, modulation, interpolation, and pulse-shape filtering according to ISO 18000-63.
EvaluateReaderTransmitChain()	Evaluates several aspects of the ISO 18000–63 transmit baseband logic and exemplifies transmit spectrum mask analysis.
GenRandFM0Frame()	Generates transponder FM0 waveforms containing the pilot tone and preamble followed by random data with prescribed SNR.
GenRandQPSKFrame()	Generates QPSK waveforms with random data and prescribed SNR.
ZCrossFM0Demod()	Performs FM0 signal demodulation using a simple zero-crossing detection algorithm.
DPLLMnMClockRecovery()	Recovers FM0 clock using DPLL and Mueller and Muller timing error estimation.
SampleRateFM0FrameSync()	Detects FM0 preamble in the sample space using cross-correlation.
SymbolRateFM0FrameSync()	Detects FM0 preamble in the symbol space using an efficient recursive algorithm.
FM0SymbolDemap()	Demaps FM0 symbols and retrieves the transponder binary data stream.
EvaluateTransponder2ReaderUplinkChain()	Evaluates ISO 18000–63 transponder to reader uplink baseband logic.

DSB-ASK and PR-ASK (Figures 5.6–5.8). Note that the SSB-ASK-modulated spectrum is not centered around the carrier and a frequency offset must be applied to the spectrum integration to recenter the spectrum mask (see the example in Figure 12 of [10]).

2. Adapt the provided code to perform spectrum mask analysis for the ETSI masks defined in [28] for UHF RFID readers. Refer to [29] for an example of ETSI mask analysis for a WLAN system.
3. To account for spectrum regrowth introduced by the transmit power amplifier, include the necessary code to model the non-linearity of the power amplifier (for details on how to use MATLAB to model non-linearities, refer to [29] and [30]).

5.7.2   Mueller and Muller DPLL Algorithm for Quadrature Reception

Make the required modifications to the code provided in *DPLLMnMClockRecovery()* to handle symbol synchronization of complex transponder baseband signals. Use (5.4) to compute the Mueller and Muller timing error and the function *GenRandFM0Frame()* to generate complex transponder baseband waveforms for testing your implementation.

5.8   Complementary Reading

1. T. F. Collins, R. Getz, D. Pu, and A. M. Wyglinski, *Software-Defined Radio for Engineers*, Artech House, 2018.
2. F. Zheng and T. Kaiser, *Digital Signal Processing for RFID*, 1st edition, Wiley, 2016.

References

[1] ISO/IEC "18000-63:2013 Information Technology – Radio Frequency Identification for Item Management – Part 63: Parameters for Air Interface Communications at 860 MHz to 960 MHz Type C," 2013.

[2] EPCglobal Inc. "EPC Radio-Frequency Identity Protocols Generation-2 UHF RFID Specification for RFID Air Interface Protocol for Communications at 860 MHz–960 MHz (version 2.0.0 ratified)," 2013.

[3] Impinj, "Indy R2000 UHF Gen 2 RFID Reader Chip." [Online], available at: www.impinj.com

[4] T. F. Collins, R. Getz, D. Pu, and A. M. Wyglinski, Software-Defined Radio for Engineers. Artech House, London, 2018.

[5] National Instruments, "Pulse-Shape Filtering in Communications Systems." [Online], available at: www.ni.com/en-us/innovations/white-papers/06/pulse-shape-filtering-in-communications-systems.html

[6] M. Viswanathan, *Wireless Communication Systems in Matlab*, 2nd ed. 2020.

[7] MathWorks, "Single Sideband Modulation via Hilbert Transform." [Online], available at: www.mathworks.com/help/signal/examples/single-sideband-modulation-via-the-hilbert-transform.html

[8] R. Kuc, *Introduction to Digital Signal Processing*. BS Publications, Hyderabad, 2008.

[9] D. M. Dobkin, *The RF in RFID: Passive UHF in Practice*, 2nd ed. Newnes, Oxford, 2012.

[10] Z. Wang et al., "Anti-Collision Scheme Analysis of RFID System," Auto-ID Labs White Paper WP-HARDWARE-045, 2007.

[11] J. Kimionis and M. M. Tentzeris, "Pulse Shaping: The Missing Piece of Backscatter Radio and RFID," *IEEE Transactions on Microwave Theory and Techniques*, 64 (12):4774–4788, 2016.

[12] R. Correia et al., "Chirp Based Backscatter Modulation," in *2019 IEEE MTT-S International Microwave Symposium (IMS)*, 2019, pp. 279–282.

[13] F. Zheng and T. Kaiser, *Digital Signal Processing for RFID*, 1st ed. Wiley, Chichester, 2016.

[14] B. B. Purkayastha and K. K. Sarma, *A Digital Phase Locked Loop based Signal and Symbol Recovery System for Wireless Channel*. Springer, New York, 2014.

[15] X. Zhang and M. Tentzeris, "Applications of Fast-Moving RFID Tags in High-Speed Railway Systems," *International Journal of Engineering Business Management*, 3(1), 2011.

[16] C. Huang and H. Min, "A New Method of Synchronization for RFID Digital Receivers," in *Eighth International Conference on Solid-State and Integrated Circuit Technology Proceedings*, 2006, pp. 1595–1597.

[17] B. Sklar, *Digital Communications: Fundamentals and Applications*. Prentice-Hall, Hoboken, NJ, 1988.

[18] F. Gardner, "A BPSK/QPSK Timing Error Detector for Sampled Receivers," *IEEE Transactions on Communications* 34(5):423429, 1986.

[19] K. Mueller and M. Muller, "Timing Recovery in Digital Synchronous Data Receivers," *IEEE Transactions on Communications*, 24(5):516–531, 1976.

[20] G. R. Danesfahani and T. G. Jeans, "Optimisation of Modified Mueller and Muller Algorithm," *Electronics Letters*, 31(13):1032–1033, 1995.

[21] MathWorks, " Symbol Synchronization Overview." [Online], available at: www .mathworks.com/help/comm/ref/comm.symbolsynchronizer-system-object .html#bumv1zd

[22] K. C. Howland, "Signal Detection and Frame Synchronization of Multiple Wireless Networking Waveforms." Master's thesis, Naval Postgraduate School, CA, 2007.

[23] L. Xi and S. H. Cho, "A RFID Decoder Using a Matched Filter for Compensation of the Frequency Variation," in *Fifth International Conference on Wireless Communications, Networking and Mobile Computing*, 2009, pp. 1–5.

[24] D. De Donno, F. Ricciato, and L. Tarricone, "Listening to Tags: Uplink RFID Measurements with an Open-Source Software-Defined Radio Tool," *IEEE Transactions on Instrumentation and Measurement*, 62(1):109–118, 2013.

[25] MathWorks, "AWGN Channel Documentation." [Online], available at: www.mathworks .com/help/comm/ug/awgn-channel.html

[26] F. J. Harris and M. Rice, "Multirate Digital Filters for Symbol Timing Synchronization in Software-Defined Radios," *IEEE Journal on Selected Areas in Communications*, 19 (12):2346–2357, 2001.

[27] Y. Li, H. Wu, and Y. Zeng, "Signal Coding in Physical Layer Separation for RFID Tag Collision," in *International Conference on Wireless Communication, Network and Multimedia Engineering*, 2019, pp. 8–13.

[28] ETSI, "Radio Frequency Identification Equipment Operating in the Band 865 MHz to 868 MHz with Power Levels up to 2 W and in the Band 915 MHz to 921 MHz with Power Levels up to 4 W; Harmonised Standard for Access to Radio Spectrum," EN 302 208 C3.3.0, 2020. [Online], available at: www.etsi.org/deliver/etsi_en/302200_302299/ 302208/03.03.00_20/en_302208v030300a.pdf

[29] MathWorks, "802.11ad Transmitter Spectral Emission Mask Testing." [Online], available at: www.mathworks.com/help/wlan/examples/802-11ad-transmitter-spectral-emission-mask-testing.html

[30] Y. Han and H. Min, "System Modeling and Simulation of RFID," Auto-ID WP-HARDWARE-010, 2004.

6 A Simple Low-Cost RFID Reader Implementation

6.1 Introduction

In this chapter, we explore low-cost and low-complexity techniques for the design of a UHF RFID reader compliant with ISO 18000-63 [1, 2]. These techniques are suitable for applications where flexibility and performance are not major requirements. The proposed receiver implementation is based on a quadrature (IQ) data slicing demodulator and leverages special features of the FM0 encoding defined in the ISO 18000-63 standard and built-in hardware capabilities of a standard microcontroller to implement a simple yet robust FM0 decoder. We also explore a new approach to detecting errors in the received transponder data even prior to applying CRC by exploiting the memory properties of the FM0 encoding scheme.

For the transmitter, we propose a simplified implementation of the PR-ASK modulation defined in ISO 18000-63 to reduce the reader transmit bandwidth by 50% while keeping the same bit rate. Compliance with the transmit spectral mask is achieved by using an analog filter at the MCU transmit baseband output port. To demonstrate and validate the proposed concepts, we assemble an experimental RFID reader, discuss its hardware and algorithmic implementation, and present simulation and measurement results. We also provide source code in the C language and sample data to evaluate the proposed algorithms (see www.cambridge.org/9781108489713).

6.2 Proposed RFID Reader Architecture

A simplified block diagram of the proposed RFID reader is depicted in Figure 6.1. The receiver comprises an IQ down-converter mixer followed by an IQ data-slicing demodulator, an IQ combiner, and an event capture unit (part of the MCU). To further reduce complexity (at the expense of performance), the IQ mixer could be replaced with a simple envelope detector. For the transmitter, we propose a simple implementation of phase-reversal ASK modulation that uses a differential up-converter mixer driven with a three-level two-bit baseband code generated by the MCU. This scheme, detailed in Section 6.2.2, allows us to implement ISO 18000-63 phase-reversal ASK modulation without the need for a high-resolution DAC. In this

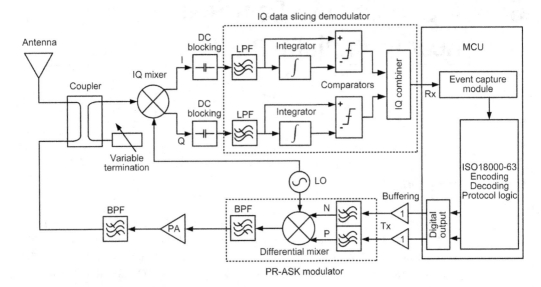

Figure 6.1. Simplified diagram of the proposed RFID reader.

design, we use a monostatic antenna configuration where the transmitting and receiving signal paths are isolated via a directional coupler/circulator. To further improve transmitter-to-receiver isolation, the isolated port of the coupler is terminated with a variable-load impedance that can be adjusted for optimal self-jamming suppression (refer to Chapter 8 for a detailed discussion about self-jamming suppression in backscatter radio systems).

6.2.1   Transmit Baseband Encoding

Figure 6.2 presents a flowchart of the proposed ISO 18000-63 PIE encoder. In this implementation, a two-level PIE baseband signal is generated through a digital output pin of the MCU and clocked by an MCU timer. After low-pass-filtering, this signal is used to drive the IF input of an up-converter mixer to produce ASK modulation. With minor changes, the algorithm presented in Figure 6.2 can be used to generate a two-bit code to drive the phase-reversal ASK modulator proposed in the next section. In this case, two digital I/O channels/pins of the MCU are required to output a phase-reversal ASK code (see Figure 6.3).

As discussed in the previous chapter, transmit pulse-shaping is needed to comply with spectrum masks imposed by regulations. This typically requires expensive high-resolution DACs. Here, we used a simple low-pass LC circuit to filter the two-level signal generated by a digital output pin of the MCU for compliance with transmit masks at a reduced design complexity. To properly drive the up-converter mixer IF input, we use a unit-gain operational amplifier buffer to enhance the MCU output current drive.

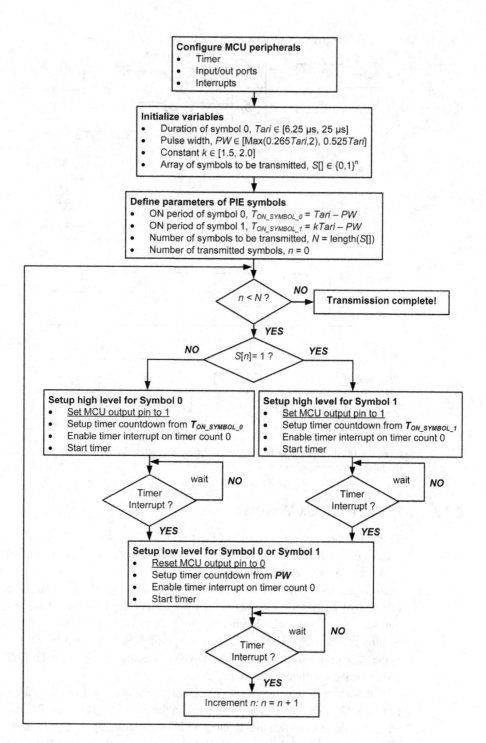

Figure 6.2 Flowchart of the proposed ISO 18000-63 PIE encoder. The parameters *PW* and *Tari* are as defined in the previous chapter.

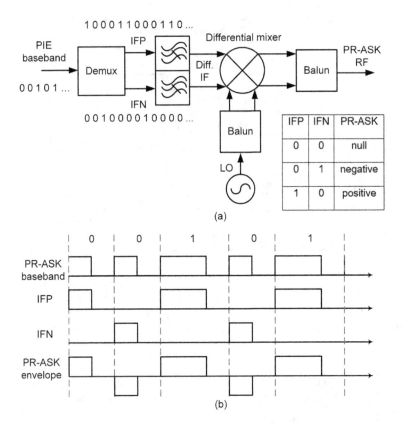

Figure 6.3 (a) Illustration of the proposed PR-ASK modulation scheme. (b) PR-ASK modulator waveforms.

6.2.2   A Simple PR-ASK Modulator

To reduce the spectrum content of the transmit RF signal, the PR-ASK modulation defined under the ISO 18000-63 standard imposes a phase inversion in the RF carrier at the beginning of every other baseband symbol (see previous chapter). This leads to an average bandwidth reduction of 50% while maintaining the same bit rate compared to standard DSB-ASK modulation. To implement the PR-ASK phase inversion, a multi-level baseband signal generated by a high-resolution n-bit DAC is typically required. Here, we propose a simple new approach to implement the phase inversion in the RF carrier using two single-bit DACs to create a two-bit three-level differential code to drive a differential mixer.

A minimum of three amplitude levels (negative, zero, and positive) are needed to generate PR-ASK modulation. To produce a three-level signal at the RF output of the mixer, we drive its differential IF input with a differential signal generated from a two-bit code derived from the original PIE data stream such that: (i) the IF input code '01' produces a negative baseband envelope that phase-reverses the RF output; (ii) the IF input code '00' produces a null RF output; and (iii) the IF input code '10' creates a

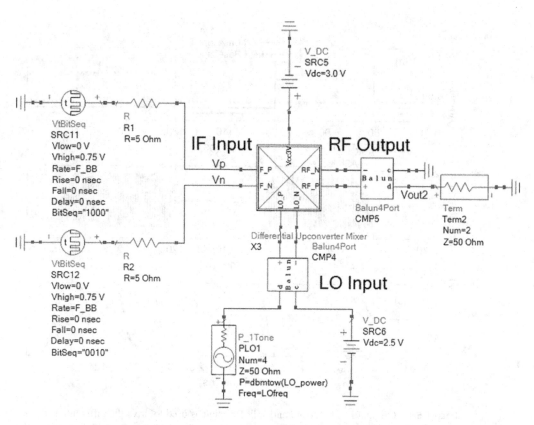

Figure 6.4 Schematic in ADS of proposed PR-ASK modulator.

positive baseband envelope for a non-reversed RF output (see Figure 6.3). The IFP and IFN sub-streams at the output of the demuxer in Figure 6.3 used to drive differential IF ports of the mixer have an average frequency equal to half the original PIE symbol rate.

To demonstrate the proposed scheme, we used Keysight's Advanced Design System (ADS) to simulate the circuit shown in Figure 6.4, where the mixer is a heterojunction bipolar transistor (HBT)-based Gilbert cell differential mixer from the ADS RFIC examples library. For illustration, we considered a baseband sequence of PIE-encoded zeros obtained by cycling the binary pattern '1010' (the dash-dotted line in Figure 6.5(a)). From this pattern, we produce the sub-stream '1000' to drive the mixer positive IF input (the solid line in Figure 6.5(a)) and the sub-stream '0010' to drive the mixer negative IF input (the dashed line in Figure 6.5(a)). The dash-dotted line in Figure 6.5(b) depicts the generated PR-ASK-modulated signal, and Figure 6.5 (c) shows the detail of the phase inversion in the generated PR-ASK RF waveform. For comparison, we also show the ASK-modulated signal (the solid line in Figure 6.5 (b)). Figure 6.6 presents the spectra of the ASK- and PR-ASK-modulated signals for a sequence of PIE zeros (square wave). Notice a 50% reduction (from 50 MHz to 25 MHz) in the PR-ASK bandwidth compared to ASK.

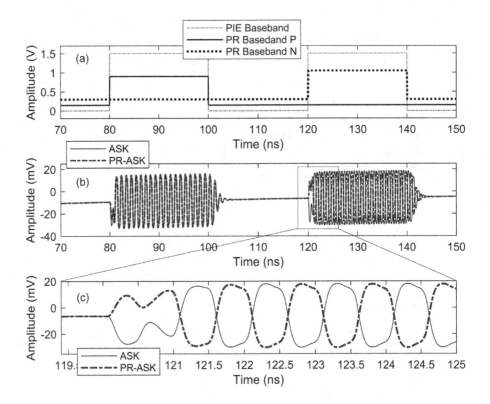

Figure 6.5 (a) Original PIE baseband and split PIE signals used to drive the differential mixer. The split signals have an amplitude offset for clarity. (b) Mixer RF output produced by ASK and PR-ASK modulation. (c) Detail of the phase inversion produced by the proposed PR-ASK modulator.

6.2.3   Quadrature Data-Slicing Demodulation

The proposed receiver is based on the quadrature[1] data-slicing demodulator depicted in Figure 6.7(a), which is an advanced version of the ASK data-slicing demodulator discussed in [3, 4]. This receiver employs two comparator circuits that operate as one-bit ADCs and are used to digitize the complex baseband signal at the output of the IQ down-converter mixer (see Figure 6.1).

A low-pass filter is used at the input of each comparator to filter out noise in the incoming IQ baseband signal, and integrator filters extract the complex average of the IQ baseband for the decision threshold in the comparator. Compared to conventional data-slicing demodulation, the proposed IQ scheme offers improved performance and prevents transponder detection blind spots that can occur when the intercepted back-scattered signal and the reader local oscillator are orthogonal / out of phase. After being digitized, the complex IQ signal is converted into a real signal by the IQ

---

[1] For the experiments conducted in this chapter, we implemented only in-phase demodulation but in general, IQ demodulation is required to prevent blind spots in transponder detection (see Chapter 8).

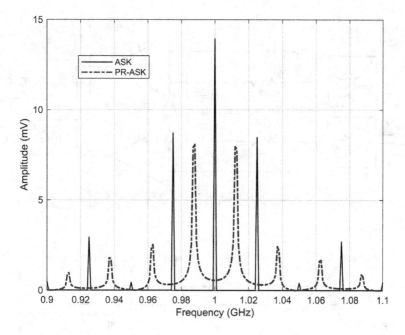

Figure 6.6 Spectra of the signals in Figure 6.5(b) showing a 50% bandwidth reduction when using PR-ASK modulation. For demonstration purposes, a periodic square wave signal is used here.

combiner (implemented with an OR gate), and then delivered to an MCU event capture unit for further processing.

6.2.4   A Simple FM0 Decoder

Event capture is typically used in embedded systems for timing external events. Here, we use the event capture hardware resource (eCAP) of the TMS320F28335 MCU to detect transitions in the FM0 signal at the output of the one-bit digitizer and record the time elapsed between these transitions. The eCAP is based on a 32-bit counter that provides a time resolution of 6.67 ns for a 150 MHz clock frequency. The eCAP operation is illustrated in Figure 6.7(b). We set the event capture unit for continuous mode operation and store the time elapsed between transitions ($t1, t2, t3, \ldots$) in its circular buffer. Once the buffer is full (which is signaled by a hardware interrupt), we collect the eCAP buffer data and save it in a software buffer. To decode the transponder message from the array of elapsed times ($t1, t2, t3, \ldots$), we use the following procedure:

1. We defined the nominal duration of
   - half the FM0 symbol '0' as $T_{0\text{Nominal}}$,
   - the full FM0 symbol '1' as $T_{1\text{Nominal}}$,
   - the full FM0 violation symbol 'V' as $T_{V\text{Nominal}}$.

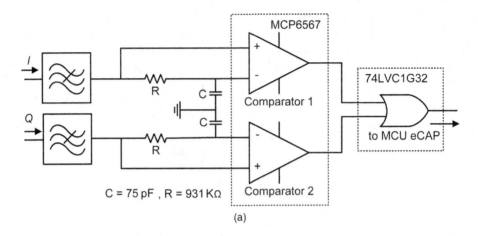

(a)

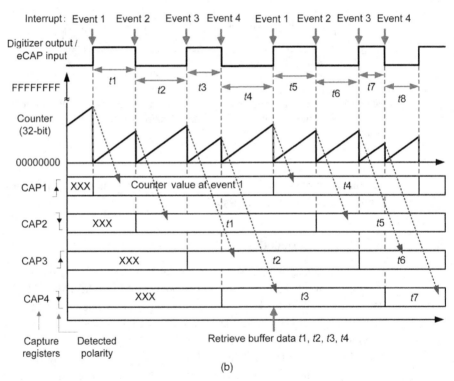

(b)

Figure 6.7 (a) IQ data-slicing demodulator, where two comparators digitize the IQ signal producing ASK modulation and an IQ combiner (OR gate) converts the digitized IQ signal into a real signal for the event capture unit. (b) Illustration of the TMS320F28335 event capture operation. (c) State machine of the proposed FM0 data decoder with code error detection capability.

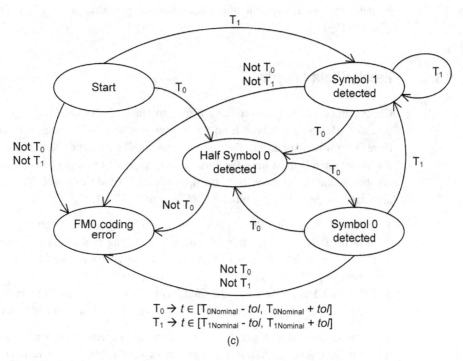

$$T_0 \rightarrow t \in [T_{0\text{Nominal}} - tol, T_{0\text{Nominal}} + tol]$$
$$T_1 \rightarrow t \in [T_{1\text{Nominal}} - tol, T_{1\text{Nominal}} + tol]$$

(c)

Figure 6.7 (cont.)

2. We define the duration range of
   - half the FM0 symbol '0' as T_0 in the interval $[T_{0\text{Nominal}} - tol, T_{0\text{Nominal}} + tol]$, with the parameter *tol* accounting for deviations in the nominal value of the backscatter link frequency (BLF), which can be as large as ±22% [1, 2],
   - the full FM0 symbol '1' as T_1 in the interval $[T_{1\text{Nominal}} - tol, T_{1\text{Nominal}} + tol]$,
   - the full FM0 violation symbol 'V' as T_V in the interval $[T_{V\text{Nominal}} - tol, T_{V\text{Nominal}} + tol]$.
3. In evaluating the elements of the array of elapsed times between transitions $(t1, t2, t3, \ldots)$:
   - if two consecutive elements of duration T_0 are found, we assign a logic symbol '0';
   - if an element of duration T_1 is detected, we assign a logic symbol '1';
   - if an element of duration T_V is found, we assign an FM0 preamble violation symbol 'V'.

In addition, we leverage the memory feature of the FM0 encoding scheme to detect FM0 encoding errors even prior to CRC verification. This can save processing time since the CRC verification can be skipped if errors are detected during the decoding phase. An FM0 encoding error could be, for instance, the appearance of an odd number of transitions of duration T_0 followed by a transition of duration T_1 or transitions not complying with T_0 or T_1. The state diagram of the proposed FM0 decoder with error detection capability is depicted in Figure 6.7(c). Code in the

C language to implement this algorithm is provided at www.cambridge.org/9781108489713.

6.3   Receive Data Integrity

Checksum generation and verification are important parts of the RFID communication protocol. Messages received from the transponder should be checked to ensure that there were no errors during the communication. On the other hand, CRC-protected reader commands must have the correct CRC appended to them or will otherwise be ignored by the transponder. Both the *AppendCRC5()* and *ComputeCRC16()* functions are based on the following bitwise CRC algorithm [5–7]:

1. Initialize the 16-bit CRC register with the 16-bit CRC preset value.
2. Shift the CRC register left by one bit while shifting in the next message bit.
3. If the bit just shifted out is 1, XOR the CRC register with the CRC polynomial.
4. Repeat steps 2 and 3 until there are no message bits left.
5. XOR the CRC register with the final XOR value. For a final XOR value equal to 0xFFFF, this operation corresponds to applying a 1's complement to the CRC.

The bitwise implementation can become inefficient, especially for long messages. The function *ComputeCRC16TableMethod()* offers an efficient alternative to compute the CRC-16 based on the following bytewise table-based CRC algorithm from [5, 6]:

1. Compute 256 CRC words and store in a table.
2. Initialize the 16-bit CRC register with the 16-bit CRC preset value.
3. XOR the CRC most significant byte with the incoming message byte.
4. Use this byte to index into the pre-computed 256-entry table.
5. Shift the CRC register to the left by one byte.
6. XOR the CRC register with the value indexed into the table.
7. Repeat steps 3 to 6 until no more message bytes are left.
8. XOR the CRC register with the final XOR value.

The CRC-16 table can be pre-computed by the function *CRC16BuildTable()* and stored in the FLASH memory of the reader CPU to minimize processing time.

6.4   Experimental Testbed and Results

This section describes the testbed used to study the concepts proposed in this chapter and presents experimental results, including reader waveforms and transponder EPC IDs.

6.4.1   Experimental Testbed

Figure 6.8 shows a simplified diagram of the RFID reader and experimental testbed used in this work. The reader local oscillator was implemented using Analog Devices'

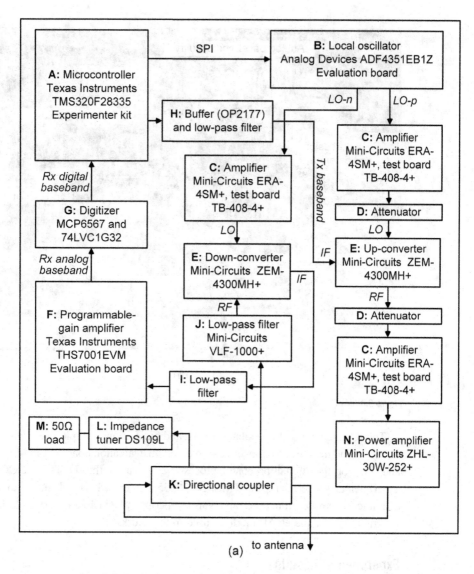

Figure 6.8 (a) Simplified diagram of the RFID reader hardware used in this work. (b) Photograph of the experimental testbed. Self-jamming suppression is described in detail in Chapter 8.

frequency synthesizer ADF4351. Its evaluation board EVAL-ADF4351EB1Z incorporates a fractional-N PLL, voltage-controlled oscillator (VCO), temperature-compensated crystal oscillator (TCXO) phase reference, and PLL loop filter.

The EVAL-ADF4351EB1Z can be interfaced using its software interface ADF435xPLL via a universal serial bus (USB), but for the experiments conducted in this chapter we used a dedicated MCU to control the synthesizer via the serial peripheral interface (SPI). To drive the LO inputs of our single-ended up-converter and down-converter mixers, we used the differential outputs of the ADF4351 RF port.

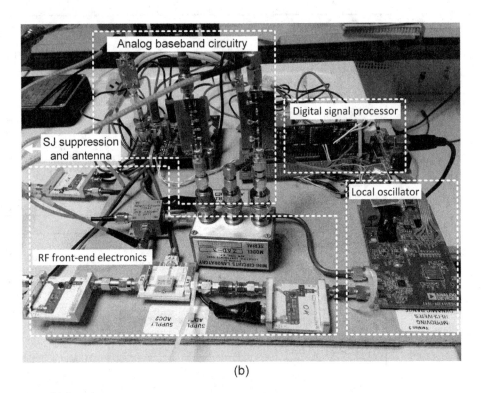

(b)

Figure 6.8 (cont.)

The IQ data-slicing demodulator, filters, and bidirectional coupler were custom made in-house, and the other components were based on evaluation boards from Mini-Circuits and TI. For the digital baseband processing, we used the TI TMDSDOCK28335 experimenter kit featuring the TMS320F28335 Delfino MCU, which combines DSP and microcontroller functions and various peripherals including the SPI, I/O port, Interrupt, Timer, and eCAP modules used in this work.

6.4.2   Experimental Results

Figure 6.9 shows the time-domain waveforms and spectra of PIE-encoded, DSB-ASK-modulated signals at the output of the power amplifier (block N). The baseband PIE signal was created using an encoding algorithm like the one in Figure 6.2 and generated via a digital output pin of the MCU.

To shape the transmit baseband pulses for compliance with transmit masks, a simple second-order LC low-pass filter can be used at the digital output pin of the MCU. For the experiments conducted in this chapter, we used a center carrier frequency of 866 MHz and a *Tari* of 6.25 µs.

Figure 6.10 shows the input and output of the data-slicing demodulator/digitizer. The analog input signal is an amplified version of the incoming baseband signal at the output of the down-converter mixer. After low-pass filtering, this signal was digitized

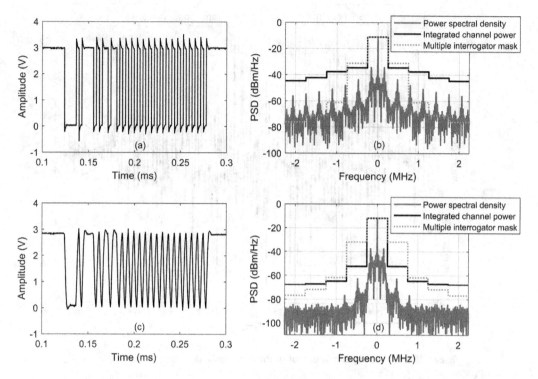

Figure 6.9 PIE-encoded, DSB-ASK-modulated signal at the output of the PA on a 866 MHz carrier and *Tari* of 6.25 μs. (a) and (b): Unfiltered transmit waveform and spectrum. (c) and (d): Filtered transmit waveform and spectrum. The applied filter was effective in meeting the multiple interrogator mask in the first and second adjacent channels, but further optimization is needed for the third and fourth channels.

and fed into the MCU event capture unit for further processing. Figure 6.11 shows elapsed times (in CPU clock cycles) measured by the event capture scheme of Figure 6.7(a) and (b) for typical transponder RN16 and EPC responses. The transponder transmission began with a 12 FM0 zero pilot tone followed by an FM0 preamble and transponder payload data.

From Figure 6.11, we can see that a half-period of symbol '0' ($T_0/2$), one period of symbol '1' (T_1), and one period of the FM0 violation symbol (T_V) spanned 506, 934, and 1452 CPU clock cycles, respectively, which corresponds to $T_0/2 \approx 3.37$ μs, $T_1 \approx 6.23$ μs, and $T_V \approx 9.68$ μs for a CPU clock frequency of 150 MHz. Symbol '0' presents roughly the same period as symbol '1', and the violation symbol has approximately 3/2 the duration of symbols '0' and '1', which complies with FM0 encoding. The average BLF of the three symbols (154 kbps) approximates the target BLF of 160 kbps commanded to the transponder by the *Query* command. The transponder RN16 and EPC were decoded as 0011000011000111 (binary) and E2003411B802011111365086 (hexadecimal) with a corresponding CRC-16 of 615D (hexadecimal), respectively.

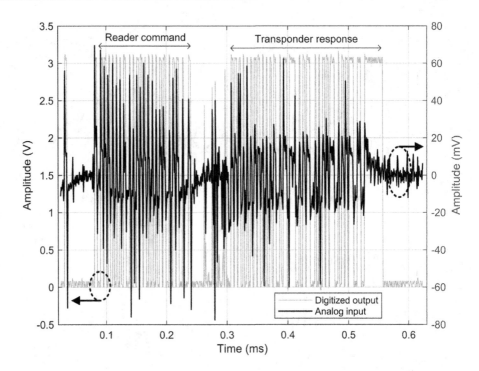

Figure 6.10 In-phase input and output of the digitizer/data-slicing demodulator.

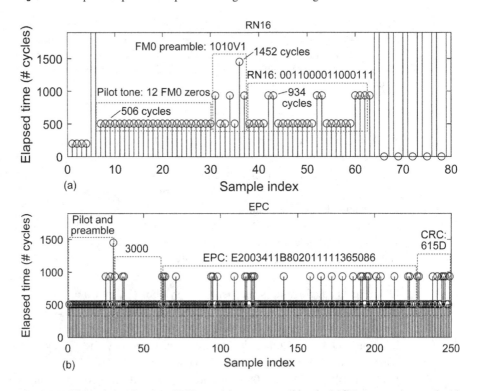

Figure 6.11 Elapsed time between FM0 transitions measured by the MCU event capture unit. (a) Transponder RN16 response. (b) Transponder EPC response. Elapsed times are given in CPU clock cycles for a CPU clock frequency of 150 MHz.

6.5   Source Code

High-level source code in the C language for generating reader commands and decoding transponder responses, including CRC checksum generation and verification and sample data for evaluating the provided code, are available at www.cambridge.org/9781108489713. The provided code can be seamlessly integrated into custom designs.

For compiling and debugging the source code provided with this book, we used the integrated development environment (IDE) from Texas Instruments (TI), Code Composer Studio version 5 (CCSv5.5.0) in simulation mode.[2] The code provided in this chapter is summarized in Table 6.1.

6.6   Conclusions

In this chapter, we explored simple and low-cost techniques for the design of a UHF RFID reader compliant with ISO 18000–63. Getting rid of ADCs and DACs substantially lowers the design cost and complexity at the expense of performance and flexibility. Cost and complexity could be further lowered at the expense of performance, for instance by using a low-cost transmitter (e.g., the TH72035 [8]) and MCU (e.g., the MSP430). To determine the maximum achievable communication range, we used gain amplifiers and an external power amplifier [9]. Depending on the desired coverage range, some of these components may not be required.

Despite its simplicity, the proposed receiver (IQ data-slicing demodulator and FM0 decoder) proved very robust at very low signal-to-noise ratio. The constructed reader achieved a coverage range of up to 9 m for an average transmitted power level of 31.5 dBm, a reader-to-transponder *Tari* of 6.25 µs, and a transponder-to-reader data rate (BLF) of 160 kbps. The FM0 demodulation could alternatively be implemented using a general timer/interrupt of the MCU (see Section 6.7.2), and these techniques can also be applied to decode Miller-encoded signals. For advanced demodulation techniques involving, for example, matched filtering and maximum likelihood detection, refer to Chapters 5 and 7.

6.7   Exercises

This section proposes experiments relating to the techniques previously discussed in this chapter involving PR-ASK modulation, FM0 decoding, and CRC check.

---

[2] Note: At the time this book was written, TI CCS IDE versions newer than 5 did not support simulation mode.

Table 6.1 Summary of signal processing functions implemented in this chapter.

Function	Description
ComposeReaderCommand()	This function composes the reader command binary sequence. The CRC is calculated and appended to the command binary sequence later.
DetectFM0PilotTone1()	This function detects the transponder FM0 pilot tone.
DetectFM0Preamble1()	This function searches for the FM0 preamble pattern and returns the initial transponder data index in the array.
DecodeFM0Data1()	This function implements the FM0 decoder illustrated in Figure 6.7(c). By exploiting the memory feature of the ISO 18000-63 FM0 encoding, this decoder can detect encoding errors prior to CRC verification.
AppendCRC5()	This function computes the ISO 18000–63 5-bit CRC checksum of a message and appends it to the message.
ComputeCRC16()	This function computes the ISO 18000–63 16-bit CRC checksum of a message.
CheckCRC16()	Using the function *ComputeCRC16()*, this function verifies the validity of a message against its appended 16-bit CRC checksum.
AppendCRC16()	Using the function *ComputeCRC16()*, this function computes the 16-bit CRC checksum of a message and appends it to the message.
ComputeCRC16TableMethod()	This function performs a 16-bit CRC computation based on the bytewise table-based CRC algorithm discussed in [5, 6].
Main()	This script exemplifies the use of the provided source code. Sample data for code evaluation are provided in the header file SampleData.h.

6.7.1   PR-ASK Modulation

Implement the modifications needed in the algorithm presented in Figure 6.2 to generate a two-bit differential code for driving the PR-ASK differential mixer scheme proposed in Figure 6.3. Experimentally evaluate the proposed PR-ASK modulation scheme using your two-bit code and a differential mixer.

6.7.2   Alternative FM0 Demodulation Scheme

An alternative method for detecting FM0 or Miller transitions based on a general-purpose MCU timer and interrupt can be implemented as follows:

1. Define the maximum number of transitions allowed, M, and initialize the number of detected transitions, $m = 0$.
2. Configure your MCU to detect external rising and falling edge interrupts on the input pin that your FM0 or Miller-encoded baseband signal is connected to.
3. Configure your MCU timer for countdown mode.
4. Enable the external pin interrupt and start the timer countdown from its maximum period, T_{init} (e.g., for a 32-bit counter timer, $T_{init} = 0xFFFFFFFF$).
5. If an interrupt is detected, proceed to 6. Otherwise, keep waiting for an interrupt.
6. Disable external interrupt, stop timer, retrieve final timer counter value, T_{final}.

7. Save elapsed time, $T_{elapsed}[m] = T_{init} - T_{final}$, and increment number of detected transitions, $m = m + 1$.
8. Repeat steps 4 to 7 while $m < M$.
9. Use the functions provided in Section 6.5 to further process the data stored in $T_{elapsed}[]$ and decode the transponder message.

6.7.3 Efficient CRC Computation

The table-based bytewise CRC algorithm used in this chapter (see the code provided) is generally more efficient than the bitwise algorithm, but it only works for messages containing whole bytes. To improve the efficiency for an arbitrary number of bits *NDataBits*, implement the following hybrid algorithm based on the functions *ComputeCRC16TableMethod()* and *ComputeCRC16()*:

1. Determine the number of full bytes in the message: *NBytes* = *NDataBits*/8.
2. Determine the remainder number of bits in the message: *NBits* = *NDataBits*%8.
3. Use the function *ComputeCRC16TableMethod()* with the initial CRC preset to compute the CRC for the first *NBytes* full bytes of the message.
4. XOR the 16-bit CRC value obtained in step 3 with 0xFFFF (one's complement).
5. Use function *ComputeCRC16()* with the 16-bit value obtained in step 4 as the new CRC preset to compute the CRC for the remaining *NBits* bits of the message. This gives the final CRC for the full message.
   [/ – Integer division; % – Remainder of division]

References

[1] ISO/IEC "18000-63:2013 Information Technology – Radio Frequency Identification for Item Management – Part 63: Parameters for Air Interface Communications at 860 MHz to 960 MHz Type C," 2013.
[2] EPCglobal Inc. "EPC Radio-Frequency Identity Protocols Generation-2 UHF RFID Specification for RFID Air Interface Protocol for Communications at 860 MHz–960 MHz (version 2.0.0 ratified)," 2013.
[3] P. Nikitin, S. Ramamurthy, and R. Martinez, "Simple Low Cost UHF RFID Reader," in *Proc. 2013 IEEE International Conference on RFID*, 2013.
[4] Maxim, "Data Slicing Techniques for UHF ASK Receivers," application note. [Online], available at: www.maximintegrated.com/app-notes/index.mvp/id/3671
[5] R. N. Williams, "A Painless Guide to CRC Error Detection Algorithms," 1993.
[6] Texas instruments, "CRC Implementation with MSP430," application report SLAA221, 2018.
[7] L. Bies, "30 Years Experience in One Website." [Online], available at: www.lammertbies.nl
[8] Melexis, "TH72035 Transmitter IC – 868/915 MHz, FSK/ASK." [Online], available at: www.melexis.com
[9] Mini-Circuits, "ZHL-30W-252+ Power Amplifier." [Online], available at: www .minicircuits.com

7   A Software-Defined RFID Reader Design

7.1   Brief Introduction to Software-Defined Radio

Software-defined radios (SDRs) use reconfigurable hardware and adaptable software to implement some or all radio physical layer functions to achieve multi-band, multi-modulation, and multi-standard operation [1, 2]. This radio paradigm, first introduced by Joseph Mitola [3] and initially used in the military to allow backward compatibility and interoperability, has become commonplace in modern civilian and commercial radio systems [4].

A typical SDR comprises an RF front end (including up- and down-converters) and a digital processing unit interfaced via digital-to-analog and analog-to-digital converters (see Figure 7.1). Digital processing can be implemented on a general-purpose processor (GPP), a digital signal processor (DSP), or a field-programmable gate array (FPGA). The more functions of the radio implemented via reconfigurable hardware or software, the more efficient, flexible, and versatile the SDR will be. The ideal SDR would be based on the direct RF conversion approach where RF signals are digitized directly at the antenna. This is a challenging proposition that requires fast, expensive, and energy-demanding signal conversion and processing hardware, but there has been significant progress toward a practical implementation of this concept, with the some of the most promising approaches being based on sub-sampling [1] and single-bit converters [5].

In RFID, SDR enables flexible, multi-standard, and upgradable reader designs [5–10]. A popular SDR research hardware platform is the Universal Software Radio Peripheral (USRP) developed by Ettus Research [11]. This platform is commonly paired with open-source GNU Radio software [12] or LabView [13] for RFID applications [6–8]. In [5], an all-digital SDR design based on high-speed single-bit conversion and FPGA was proposed for a multi-standard reconfigurable RFID reader. In [10], the authors tested cost-effective UHF front ends based on off-the-shelf devices for SDR RFID vehicular applications.

In the remainder of this chapter, we report on the design of an SDR-based UHF RFID reader compliant with the ISO 18000-63 standard. The analog front end is built around the AD9010 IC discussed in [10], and the baseband signal processing is implemented on a multi-core DSP. A key SDR-enabled feature of this design is its ability to generate non-conventional powering waveforms that can potentially drive RFID transponders more efficiently and increase their operational range [14, 15].

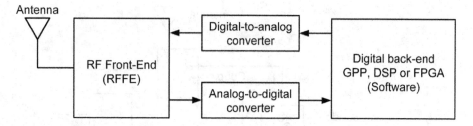

Figure 7.1 Simplified diagram of a typical SDR.

7.2   Reader Hardware Design

In this section we describe our hardware implementation, including analog/RF front end, signal converters and digital signal processor, and multicarrier wireless power transmitter.

7.2.1   Analog/RF Front End

The RFID reader design proposed here is based on a homodyne or direct conversion architecture (Figure 7.2) featuring a monostatic configuration where the transmitter and receiver share an antenna through a directional coupler. The transmit complex (I/Q) baseband signal, generated in a DSP, is converted to the analog domain by two DACs and translated to the desired RF carrier frequency in the UHF frequency range by an I/Q modulator (I/Q up-converter mixer). The modulator RF output signal is then amplified and applied to the antenna. To minimize out-of-band emissions, a bandpass filter can be applied to the output of the power amplifier (PA). Additional signal conditioning circuitry (not shown here) is required to properly drive the intermediate frequency (IF) ports of the modulator.

In the receiver, the transponder backscattered signal is sampled at the transmitter–receiver isolator and translated to baseband by the I/Q demodulator (I/Q down-converter mixer). A low-noise amplifier (LNA) can be used prior to the demodulator to improve reception performance, but this approach requires sufficient RF self-jamming suppression to prevent distortion in the LNA caused by transmitter-to-receiver leakage. A detailed discussion on leakage and self-jamming in passive-backscatter systems, including algorithm implementations, is presented in Chapter 8. If no self-jamming suppression is employed, then the ability to properly handle large RF transmitter-to-receiver leakage becomes more important than achieving exceptional noise performance, in which case the use of an LNA is not recommended. An (optional) bandpass filter can be used at the input of the demodulator to improve the receiver's frequency selectivity.

To remove noise and high-frequency signal components produced during down-conversion, the IF outputs of the IQ demodulator are low-pass filtered. The resulting

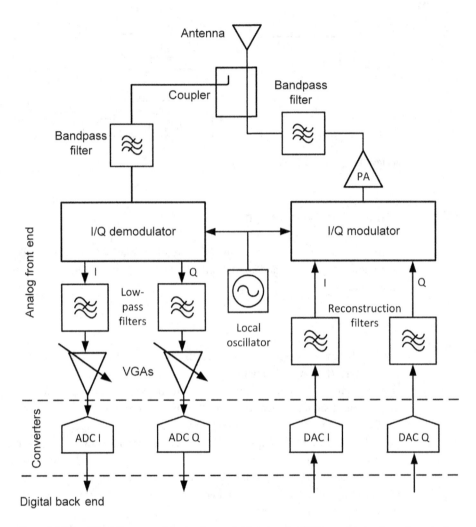

Figure 7.2 Simplified diagram of the implemented front end with transmitter and receiver sharing the same local oscillator and antenna.

complex signal then goes through DC blocking filters (not shown in Figure 7.2) to remove DC offsets created by self-jamming. Careful design of the DC blocking filter is required to prevent transients that can distort the transponder baseband signal, especially at high transponder data rates. This problem and its mitigation, including advanced filter techniques with transient compensation, are discussed in Chapter 8. The RF front end used in this chapter incorporates built-in circuitry for implementing one such technique.

The filtered and transient-compensated complex baseband signal is then amplified by built-in variable gain amplifiers (VGAs) in the IC front end and converted to the digital domain by two ADCs. The digitized signal is then processed in the DSP to decode the transponder message as described in Sections 7.3.3 and 7.3.4.

The analog front-end was built around the Analog Devices ADF9010 IC transceiver [16], which integrates analog receive baseband circuitry and an RF transmitter modulator operating in the 840–960 MHz frequency range. The I/Q receive chain incorporates baseband programmable gain amplification (PGA), baseband filtering, and signal conditioning for the ADC interface. The receiver baseband gain ranges from 3 dB to 24 dB, programmable in steps of 3 dB, and the receive low-pass filter can be programmed with a cut-off frequency of 330 kHz, 880 kHz, or 1.76 MHz; a bypass mode is also available. The transmit chain consists of an I/Q modulator followed by an output amplifier driver.

The ADF9010 also includes an on-chip integer-N frequency synthesizer and voltage-controlled oscillator (VCO), which was used in this project to generate the LO signal for the reader transmitter and receiver. But this transceiver does not offer demodulation capabilities. For I/Q demodulation, we used the ADL5382 down-conversion mixer [17] (which is rated for a frequency range from 700 MHz to 2.7 GHz) and drove it using the LO signal of the ADF9010 available externally. Configuration of the ADF9010 internal register bank was done by the DSP via the SPI.

7.2.2   DA and AD Converters

We used the AD9709 [18] and DAC5662 [19] dual-channel differential DACs for generating the baseband I/Q data and multicarrier waveforms, respectively (see Figure 7.3 and Section 7.2.5). The AD9709, which has a resolution of 8 bits and maximum sampling rate of 125 MSPS, was operated at 2 MSPS at its full resolution; the DAC5662, which has a 12-bit resolution and can support up to 275 MSPS sampling rate, was operated at 100 MSPS with only 8 bits of resolution. The AD9709 circuitry was custom integrated into our main front-end board design, whereas the DAC5662 was integrated using its evaluation board [20].

For digitizing the received baseband signal, we used the 12-bit built-in ADCs of the DSP. These ADCs have a modest sampling speed but worked relatively well for our prototype. Faster standalone ADC devices are presented in Table 7.1.

7.2.3   Digital Signal Processor

For baseband signal processing, we selected the 32-bit dual-core TMS320F28377D processor from Texas Instruments Inc. [21]. This device, which is primarily intended for sensing and control, offers an excellent cost–benefit tradeoff by combining DSP capabilities with a rich set of peripherals and built-in memory; it includes:

- 12/16-bit multi-channel ADC
- direct memory access (DMA)
- pulse-width modulation (PWM)
- universal serial bus (USB)
- 8-bit universal parallel port (UPP)

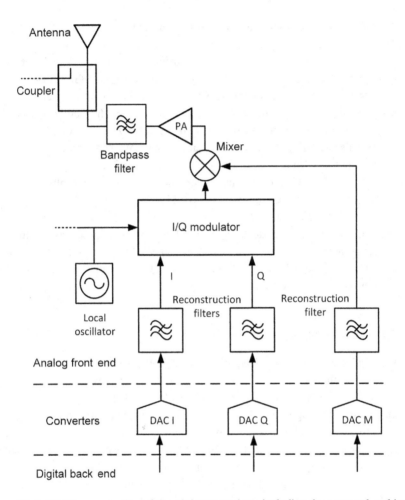

Figure 7.3 Diagram portion of the reader transmitter, including the proposed multicarrier generator. Multicarrier RFID interrogation signals are generated by mixing the usual modulated RFID signal with baseband multisine envelopes generated using a lookup-table approach (described later) via a dedicated DAC (DAC M).

- serial peripheral interface (SPI)
- 1 MB on-chip flash memory
- 204 KB on-chip RAM memory.

The main limitation for our application was the relatively low sampling rate of the built-in ADC system (3.5 MSPS maximum sampling rate per channel and up to 14 MSPS throughput when combining multiple channels). But the ability to implement analog-to-digital conversion, signal processing, and network connectivity within a single device was a major advantage as it substantially reduced the design complexity, part count, and cost. In this work, we used the TMS320F28377D DSP through its TMDSDOCK28379D experimenter board [22]. For comparable low-cost processor alternatives suitable for this application, see Table 7.2.

Table 7.1 Alternative components suitable for SDR UHF RFID reader design.

Analog front end	
Part number	Description
AD9361, AD9363, AD9364	High-performance transceivers with full transmit and receive RF and analog baseband circuitry, and ADCs/DACs. These ICs could replace the analog front end, demodulator, and ADCs/DACs used here.
ADALM-PLUTO SDR	Low-cost integrated SDR learning module based on the AD9363.
ST25RU3991/2/3	Family of UHF RFID reader transceivers, primarily intended to work as standalone readers but offer the option to bypass protocol-specific baseband circuitry and work as analog front ends for custom SDR.
ADF4351, ADF4360-7	Frequency synthesizers with integrated VCO suitable for RFID UHF LO generation. The RFID reader design presented in Chapter 6 uses ADF4351.
Modulation and demodulation	
LT5568-2, TRF37A32	Modulator and demodulator suitable for UHF RFID applications.
AD and DA conversion	
AD9709, MAX5864, MAX5188, MAX5191, DAC5662, AD9963, AD9861	Dual-channel DACs and ADCs suitable for I/Q signal conversion in UHF SDR RFID readers. The AD9963 has been evaluated in [10] in conjunction with the AD9361 transceiver for UHF RFID applications.
Baseband DSP	
	See Table 7.2 for low-cost processors that can suit UHF SDR RFID reader designs and be used as alternatives to the TMS320F28377D.

7.2.4   Other Components

Other key components used in our design include the RCP890Q10 directional coupler from RN2 Technologies [23] and the SKY6511 power amplifier from Skyworks Solutions [24]. The RCP890Q10 directional coupler features a compact design with a 10 dB coupling factor and is rated for the 815–960 MHz frequency range. The selected power amplifier covers the 800 MHz to 1000 MHz band and features an output power of 33 dBm and 1 dB compression point of 27 dBm. The bill of material also includes baluns ETC1-1-13TR and 0900BL18B100 used for differential to single-ended conversion, temperature-compensated crystal oscillator FOX924 used as phase reference for the ADF9010, and low-pass filter LFCN-1000D+ used in the transmitter and voltage regulator ADP3334ARZ. Tables 7.1 and 7.2 list alternative

Table 7.2 Low-cost DSP solutions.

	TMS320 F28377D (Used here)	TMS320 F28335 (Chapter 6)	TMS320 C28346	TMS320V C5410A	TMS320C6748	OMAP-L138	LPC4370	ADSP-BF537
Manufacturer	TI	TI	TI	TI	TI	TI	NXP	AD
Register length	32-bit	32-bit	32-bit	16-bit	32-bit	32-bit	32-bit	32-bit
Floating point	Yes	Yes	Yes	No	Yes	Yes	Yes	
CPU cores	2	1	1	1	1	2	2	1
CPU clock	200 MHz	150 MHz	300 MHz	160 MHz	456 MHz	456 MHz	204 MHz	600 MHz
Total processing power	800 MIPS	150 MIPS	300 MIPS	160 MIPS	3648 MIPS	3648 MIPS	204 MIPS	
Co-processors	2 CLAs					1 ARM	2 Cortex-M0	
On-chip FLASH	1 MB	512 KB	No	No	No	No	No	No
On-chip RAM	204 KB	64 KB	510 KB	128 KB	256 KB	256 KB	282 KB	132 KB
ADC	3.5 MSPS	12.5 MSPS	No	No	No	No	80 MSPS	No
DMA	Yes	Yes	Yes	Yes	Yes	Yes	Yes	Yes
USB	Yes	No	No	No	Yes	Yes	Yes	No
Ethernet	No	No	No	No	Yes	Yes	Yes	Yes

MSPS – Megasamples per second
MIPS – Million instructions per second
Note: This table is for informational purposes only. Depending on the specific application requirements, the reader may need to select newer and/or higher-performance devices.

parts suitable for this project, but depending on your specific application requirements you may need to select newer and/or higher-performance devices.

7.2.5   SDR-Based Multicarrier Transmitter

It has recently been shown that non-conventional signals featuring high peak-to-average power ratio (PAPR) like chaotic, ultra-wideband, and multicarrier signals can potentially improve the energy transfer efficiency of wireless power transmission systems when compared to continuous-wave (CW) signals with equivalent average power [14, 15, 25–27]. The RFID reader design presented in this chapter features multicarrier capability. In Chapter 10, we evaluate the use of multicarrier waveforms on energy-harvesting circuits, including charge pump circuits like those used in passive UHF RFID transponders.

One way of generating an RF-modulated multicarrier waveform is to multiply an RFID baseband signal by a baseband multisine signal prior to RF up-conversion. But doing such multiplication in the digital domain can inflate the required computation resources as each baseband message symbol contains many baseband multisine pulses. For the experiments conducted in this chapter, we performed the multiplication in the analog/hardware domain as illustrated in Figure 7.3 by mixing the reader-modulated RF signal generated by the main IQ DAC with a multisine baseband waveform created by a dedicated high-speed DAC (DAC M in Figure 7.3).

Desirably, the transmitted multicarrier signal should feature high PAPR at the transponder location [14, 15]. This can be achieved by imposing a constant phase progression between the multisine subcarriers as in [27]. Multicarrier backscatter modulation was introduced in Chapter 3, and an approach for retrieving transponder baseband information backscattered on a multicarrier signal will be discussed later in this chapter.

7.2.6   Multicarrier SDR Reader Prototype

Figure 7.4 depicts the proposed SDR multicarrier RFID reader annotated with the references of the main hardware component parts, and Figure 7.5 shows the partially assembled reader. Schematics and the bill of material of our custom-built analog front end (Figure 7.5(a)) are provided at www.cambridge.org/9781108489713. For the DSP and multisine DAC, we used evaluation boards from Texas Instruments Inc. The schematic and layout of the RF power amplifier (PA) are available from the manufacturer website [24]. The PA was connected between the transmit output of the analog front end (RF port J7 in Figure 5.7(a)) and the input of the reader transmit filter (RF port J3 in Figure 5.7(a)). For the multicarrier experiments conducted later in this chapter, we used a higher-power off-the-shelf PA module for efficient amplification of the transmit high-PAPR waveforms. Figure 7.5(c) depicts a custom multi-polarization antenna developed for the SDR reader. This antenna scheme comprises a suspended microstrip patch with three feed points at 0, 45, and 90 degrees, which are driven by a PIN diode-based SP3T microstrip switch (see Figure 7.5(c)). By cycling the reader

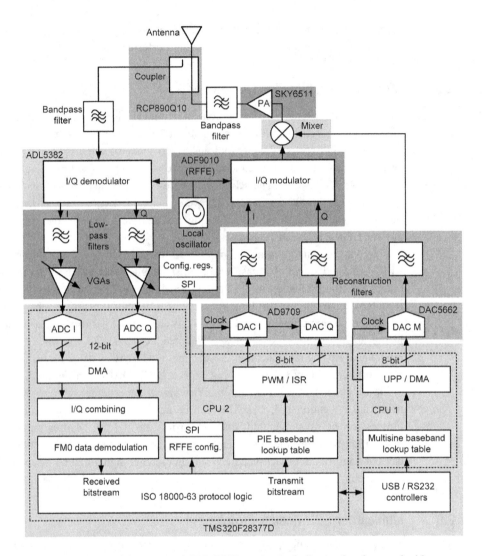

Figure 7.4 Diagram of the proposed SDR RFID reader, including analog front end with multicarrier capability and digital back end. Digital signal processing is discussed in the following sections.

transmit signal through the different polarizations, one can improve the sensitivity to polarization mismatch and multipath fading, and improve transponder reliability and coverage range.

7.3   Software Implementation

This section addresses the software for generating and transmitting reader data and power, and receiving and processing transponder data. The discussion covers CPU

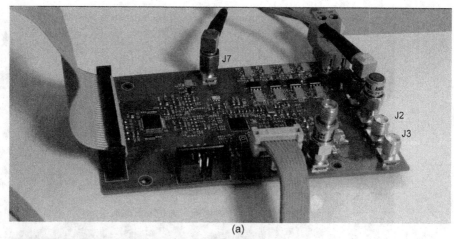

(a)

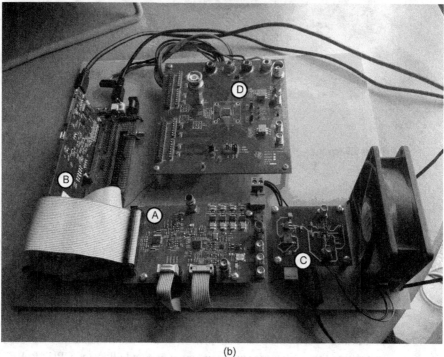

(b)

Figure 7.5 Developed hardware. (a) Custom-built analog front end. The I/Q baseband DAC (on the left side of the board) is connected to parallel ports of the DSP (left flat cable) and the analog front end IC is wired to the DSP control board via SPI and other control lines (bottom flat cables). (b) Partially assembled reader prototype comprising the custom-built analog front end (A), external PA (C), DSP (B), and multisine DAC (D). The complete reader prototype integrated with test and measurement apparatus is discussed in Section 7.4. (c) Custom UHF multi-polarized antenna scheme comprising a suspended microstrip patch antenna and a PIN diode-based SP3T microstrip switch.

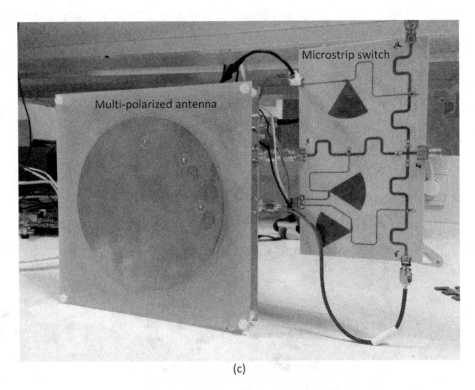

(c)

Figure 7.5 (cont.)

firmware and resources, RFID protocol logic, transponder information decoding, and signal processing optimization.

7.3.1   CPU and Firmware

The digital signal processing device used in this project (the 32-bit dual-core DSP TMS320F28377D) features two processors (CPU1, CPU2) and two co-processors (CLA1, CLA2). Each CPU/CLA runs at 200 MHz, allowing a maximum parallel processing speed of up to 800 MIPS. In this implementation, we used CPU1 for generating the multicarrier baseband envelope and allocated CPU2 to RFID baseband processing and protocol logic.

A simplified flowchart of the implemented multi-processor firmware running on CPU1 and CPU2 is depicted in Figure 7.6. Besides generating the baseband multisine envelope, CPU1, which acts as the master CPU, is responsible for booting CPU2 and initializing and allocating peripherals (e.g., input–output ports, ADC, DMA, memory, interrupts, etc.) to each CPU. Upon reset, CPU1 allocates a UPP module and a DMA channel to itself for interfacing the multisine DAC and associates two ADC channels and one PWM module to the CPU2 for digitizing the incoming RFID baseband signal and interfacing with the transmit I/Q DACs.

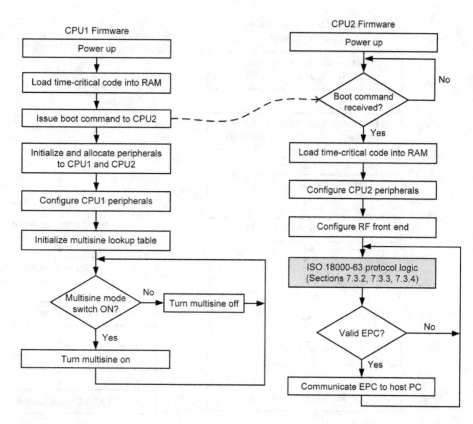

Figure 7.6 Simplified flowchart of the implemented multi-processor firmware. The baseband signal processing and ISO 18000-63 protocol logic implementation (gray rectangle) is detailed in the following sections.

Once CPU2 receives the boot command from CPU1, it first configures the RFID transmit and receive parameters (e.g., transmit frequency/power, and receive baseband gain/bandwidth) of the AD9010 RF IC front end via the SPI interface. CPU2 then processes the baseband data and implements the RFID protocol logic needed to singulate an RFID transponder, read its EPC, and access its memory (see the details in the following sections). The retrieved transponder data is then communicated to the host computer via the RS232 interface. To improve processing efficiency, both CPUs were configured to load time-critical user code from FLASH memory to RAM at boot time. A more detailed discussion of code and data processing optimization is given in Section 7.3.5.

7.3.2   Baseband Waveform Generation

To generate oversampled baseband waveforms, we used a lookup-table approach in which the desired waveform samples are synthesized in MATLAB, stored in the DSP FLASH memory, and then used in real time to update the transmit DAC (see the

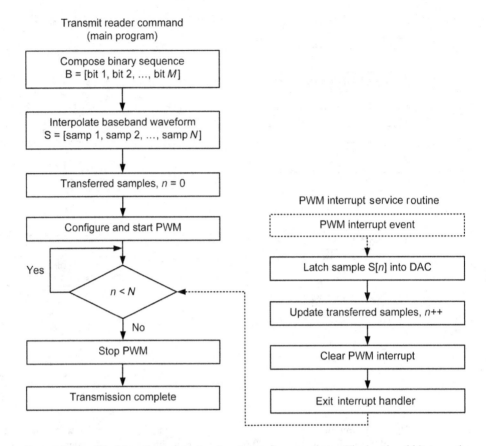

Figure 7.7 Simplified flowchart of the baseband waveform generator. The baseband bitstream is interpolated/oversampled using pre-stored data to form the transmit baseband waveform, and a PWM module and its ISR handler are used to latch the waveform samples into the transmit DAC.

simplified flowchart in Figure 7.7). This approach dramatically simplifies the computation required in the reader to generate transmit baseband waveforms.

To latch baseband waveform samples into the transmit DAC, we used a PWM module of the DSP and its associated interrupt service routine (ISR): the PWM module was configured to generate a square-wave signal through an output port of the DSP to clock the DAC. In addition, the PWM module was instructed to generate one interruption request per clock cycle, providing the timing needed to update the DAC transmit register (Figure 7.7). In our implementation, the clock frequency was set to match the desired sampling frequency of 2 MHz. The DAC eight-bit codeword for each waveform sample was obtained by normalizing and quantizing the desired analog amplitude as follows:

$$D = round\left\{\frac{255}{2}\left(\frac{A}{A_{MAX}}+1\right)\right\}, \tag{7.1}$$

Table 7.3 Lookup table examples for PIE symbols oversampled at 2 MSPS and a baseband multisine waveform oversampled at 100 MSPS. The data provided are for example purposes only; transmit pulse shapes may need to be optimized.

PIE symbol 0: [192,235,250,255,255,250,235,192,148,133,128,133,148]
PIE symbol 1: [192,235,250,255,255,255,255,255,255,255,255,255,250,235,192,148,133,128,133,148]
PIE start delimiter symbol: [255,250,235,192,148,133,128,128,128,128,128,128,128,128,128,128,128,128,128,128,128,128,133, 148]
PIE RTCal symbol: [192,235,250,255,255,255,255,255,255,255,255,255,255,255,255,255,255,255,255,255,255,255,255, 255,250,235,192,148,133,128,133,148]
PIE TRCal symbol: [192,235,250,255,255,255,255,255,255,255,255,255,255,255,255,255,255,255,255,255,255,255,255, 255,255,255,255,255,255,255,255,255,255,255,255,255,255,255,255,255,255,255,255,255,255,255,255, 255,255,255,255,255,255,255,255,255,255,255,255,255,255,255,255,255,255,255,255,255,255,255,255, 255,255,255,255,255,255,255,255,255,255,255,255,255,255,255,255,255,255,255,255,250,235,192,148, 133,128,133,148]
Baseband multisine: [255,249,231,205,175,145,121,105,99,101,110,123,134,143,146,144,137,128,120,114,112,115,121,129, 136,141,141,138,131,123,117,113,113,118,126,135,142,146,144,137,126,114,103,99,102,116,138,166,196, 224,245,254,252,237,213,183,153,127,109,100,100,107,119,131,141,146,145,139,131,122,115,112,114, 119,127,134,140,142,139,133,126,118,113,113,116,123,132,140,145,145,140,130,117,106,99,100,111,131, 158,188,217,240,253,254,242,220,192,161,134,113,101,99,105,116,128,139,145,146,141,133,124,117,113, 113,118,125,132,139,142,140,135,128,120,114,112,115,121,130,138,144,146,142,133,121,109,100,99,107, 124,149,179,209,234,250,255,247,227,200,170,141,118,103,99,102,112,125,136,144,146,143,136,127,119, 114,113,116,123,130,137,141,141,137,130,122,116,113,114,119,127,136,143,146,144,136,124,112,102,99, 104,118,142,171,201,228,247,255,250,234,208,178,149,124,107,99,101,109,121,133,142,146,144,138,130, 121,115,112,115,120,128,135,140,142,139,132,124,117,113,113,117,125,134,142,146,145,138,128,115, 104, 99,101,113,134, 162,193,221,243]

where A is the desired analog amplitude and A_{MAX} is the peak amplitude of the analog signal. An 8-bit DAC resolution allows 256 amplitude codewords or quantization levels, with code 0 corresponding to the smallest amplitude and code 255 the largest amplitude.

A similar lookup-table approach was used to generate the multisine baseband waveform. Since the multisine signal is periodic, only one period of the baseband waveform needs to be stored in memory. But the PWM method described above for generating the RFID baseband waveform was not fast enough for the multisine DAC. For the multisine generation, we used a high-speed UPP peripheral of the DSP with dedicated DMA and RAM to transfer 8-bit data to the DAC at a sampling rate of 100 MHz. Refer to [21] for details on the UPP/DMA operation.

Table 7.3 presents examples of oversampled baseband waveforms that were stored in the DSP memory for a multisine pulse and PIE '0', '1', delimiter, reader-to-transponder calibration (RTCal), and transponder-to-reader calibration (TRCal) symbols. This lookup table method can be used to generate arbitrary signals, and several baseband waveforms can be stored in memory and selected in real time. For these experiments, we generated multisine waveforms with various pulse repetition rates and different numbers of tones.

7.3.3   Digital Signal Processing

Figure 7.8 presents a simplified flowchart of the implemented reader software for the transmit and receive digital baseband processing and the ISO 18000-63 protocol logic. After issuing a transponder command, the reader begins converting the incoming IQ baseband signal through the built-in DSP 12-bit ADCs and then transfers the digitized data to a program data buffer via the DSP DMA channels. Once enough samples are buffered (determined by a user-defined threshold), the DSP starts to implement the baseband signal processing and ISO 18000-63 protocol logic.

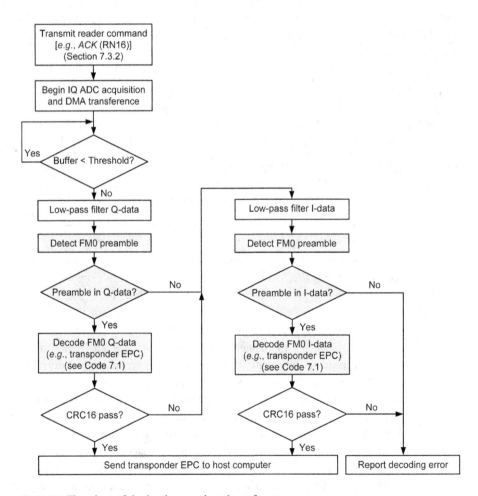

Figure 7.8 Flowchart of the implemented reader software.

For two-state BPSK transponder modulation schemes (as discussed in Chapter 5), one can drop the phase of the received complex signal by aligning the received constellation with the in-phase axis. This can be done by combining the received in-phase and quadrature components to produce a real signal as in Figure 7.4. An alternative approach consists of decoding the in-phase and quadrature data streams sequentially as in [28]. In this approach, illustrated in Figure 7.8, we first process one component of the received complex baseband signal (I or Q). If the transponder EPC/data is detected in the selected data stream, we simply skip the second data stream. If neither the preamble nor checksum is detected in the selected data stream, we process the second data stream (Q or I). For a successful in-phase or quadrature decoding, the decoded transponder EPC/data is communicated to the host computer. Otherwise, the process is terminated, and a decoding error is reported to the host computer. ISO 18000-63 RFID transponder data demodulation was discussed in Chapter 5, and a specific FM0 data demodulation procedure is described later in this chapter.

For the experiments conducted in this chapter, we used PIE and ASK modulation at the maximum bit rate of 128 kbps for reader transmissions, and we instructed transponders in the field to reply with FM0 encoding and pilot tone at data rates from 40 to 160 kbps. The singulation parameter *Q-slot* was set to zero so that the first transponder to enter the reader's interrogation field was immediately selected for subsequent transactions with the reader. You can find details of ISO 18000-63 transponder singulation and memory access in Chapter 4, and for additional reading on digital signal processing for RFID applications refer to [29–32].

7.3.4   Transponder Data Demodulation

The FM0 transponder data demodulation (the gray area in Figure 7.8) comprises two main steps: FM0 preamble detection and FM0 data decoding. Once the FM0 preamble is detected in the received transponder frame, the starting position of the payload is used for decoding the payload frame and demapping the FM0 signal into binary data. Algorithms for FM0 preamble detection and FM0 data demodulation based on zero-crossing and maximum likelihood detection, and implementation examples in MATLAB, are discussed in Chapter 5. Here, we use the zero-crossing detection method discussed in Chapter 5 for decoding transponder FM0 data. Sample code in C for an FM0 decoder based on zero-crossing detection is presented in Code 7.1.

7.3.5   Optimization

Modern RFID protocols impose stringent timing requirements for communication with transponders. For example, ISO 18000-63 defines a maximum reader reply time after which transponders will time out if they do not receive a valid reader response. Higher transponder data rates typically require faster reader processing speeds to meet these timing requirements. While this does not represent a problem for high-performance processor devices, low-cost devices with limited processing capabilities

---

Code 7.1 Simple FM0 demodulator based on zero-crossing detection

```c
typedef    unsigned int    Uint16;
#define OK 0
int16 FM0ZCDecode(Uint16 *DmaBuffer, Uint16 payStartPos, Uint16 EndPos, \
        Uint16 WaveMean, Uint16 DwnSamp, Uint16 sps, Uint16 nSymbs, Uint16
*decData)
{
    /* This function demodulates an FM0 signal using zero crossing detection.
     * The signal clock is first recovered by measuring the time (in samples)
between
     * consecutive signal transitions and the data is then decoded by
evaluating the
     * stored time durations using a predefined threshold.
     * Input:
     *      DmaBuffer - pointer to input waveform buffer
     *      payStartPos - starting position of payload data
     *      EndPos - final position in received waveform
     *      WaveMean - average amplitude of received waveform (zero, if waveform
has no DC component)
     *      DwnSamp - downsampling factor
     *      sps - number of samples per FM0 symbol
     *      nSymbs - expected number of FM0 symbols
     *  Output:
     *      decData - pointer to output binary data buffer
     * Author: Alirio Boaventura
     */
        Uint16 i = 0, m = 0, k = 0;    // Control loop variables
        Uint16 PrevSampSign = 0;       // Previous sample sign
        Uint16 CurrSampSign;           // Current sample sign
        Uint16 ElapsedTimeArray[ 400] ; // Time elapsed between zero crossings
(given in samples)
        Uint16 ElapsedTime = 0;          // Time elapsed since last zero crossing
(given in samples)
        Uint16 nZCross = 0;            // Number of zero crossings
        sps = sps/DwnSamp;             // Decimation
        Uint16 spsTol = 0.2*sps;       // sps tolerance (20%)
        // Detect zero crossings
        PrevSampSign = (DmaBuffer[ payStartPos] > WaveMean);
        for(i = payStartPos; i < EndPos; i = i + DwnSamp)
        {
            ElapsedTime++;
            CurrSampSign=(DmaBuffer[ i] > WaveMean);
            if((!PrevSampSign && CurrSampSign) || (PrevSampSign &&
!CurrSampSign)) // Detect zero crossing
            {
                ElapsedTimeArray[ nZCross] =ElapsedTime;
                nZCross++;
                ElapsedTime = 0;
            }
            PrevSampSign = CurrSampSign;
        }
        // Decode data
        m = 0;
        k = 0;
        while(m < nSymbs)
```

Code 7.1 *(cont.)*

```
        {
            // If elapsed time is within the interval [ sps-spsTol, sps+spsTol],
assert binary symbol '1'
            if( (ElapsedTimeArray[ k] > (sps-spsTol)) && (ElapsedTimeArray[ k] <
(sps+spsTol)))
            {
                decData[ m]  = 1;
                m++;
                k++;
            }
            // Otherwise assert binary symbol '0'
            else
            {
                decData[ m]  = 0;
                m++;
                k++; k++; // increment twice because symbol '0' has two
transitions
            }
        }
    return OK;
}
```

require a more efficient/optimized use of resources. Below we describe some approaches for speeding up code execution and improving the overall efficiency of the RFID reader.

- **Efficient code execution**. Firmware user code in embedded systems is typically stored in non-volatile FLASH memory. But code execution directly from FLASH memory can lead to sub-optimal performance. Modern GPPs and DSPs provide mechanisms for executing user code from RAM, allowing a significant speed-up in code execution. The DSP used in this work offers one such mechanism, where the CPU can load specific user code from FLASH to RAM at boot/power-up time and then execute the code from the RAM. We used this approach here to optimize the execution of time-critical code, for example the code used to demodulate transponder data.
- **Assembly-level optimization**. CPU-specific assembly language enables inherently more efficient coding structures that can optimize performance. This can be combined with the RAM code execution approach described above to dramatically improve execution of time-critical code.
- **Concurrent transfer and processing**. If a dedicated data transfer mechanism is available to interface with ADCs and manage the transfer of incoming transponder data, then one can use the main CPU to process the incoming data concurrently with data transfer. Here, we used the dedicated DMA channels of the DSP for transferring digitized incoming data to the DSP memory concurrently with data processing (see Figure 7.8). This approach requires a synchronization mechanism to ensure that (i) enough data are buffered before the data processing begins, and

(ii) the DMA data transfer is not overrun by the processing. Refer to [33] for further details.

- **Multitasking**. Sequential single-threaded programming can be inefficient when busy-waiting loops like those in Figures 7.7 and 7.8 are used, as valuable CPU time is wasted on useless activity. Moreover, the CPU can get stuck in infinite loops. A better alternative to single-threaded programming is to use concurrent multitasking for improved performance and efficiency.

- **Parallel processing**. When available, parallel processing can provide massive speed-ups. In this work, we used the two main CPUs of the DSP to independently implement the reader baseband processing and synthesize multicarrier baseband waveforms. For even more parallel processing power, we could also utilize the two CLAs available in this DSP.

- **Dedicated data transfer**. In this work, we used the built-in DMA and UPP modules of the DSP to interface high-speed data to the DSP memory and data converters.

- **Lookup waveform synthesis**. In this approach, we used pre-stored baseband pulses to simplify and speed up the generation of transmit baseband waveforms for both RFID reader signals and multisine envelopes (see Section 7.3.2).

7.4   Measurement Apparatus

Figure 7.9 illustrates the measurement testbed used to evaluate the RFID reader prototype in CW and multisine modes. In addition to displaying the singulated transponder EPC on a computer user interface (see Figure 7.10), we also sampled and stored waveforms at critical points of the reader transmit and receive chains to

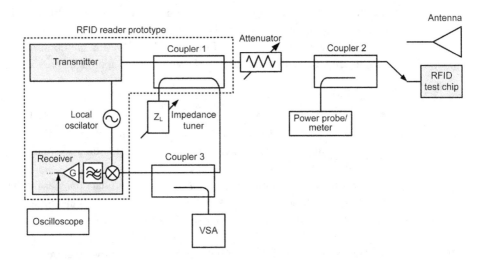

Figure 7.9 Simplified diagram of the measurement testbed used to evaluate the RFID reader prototype.

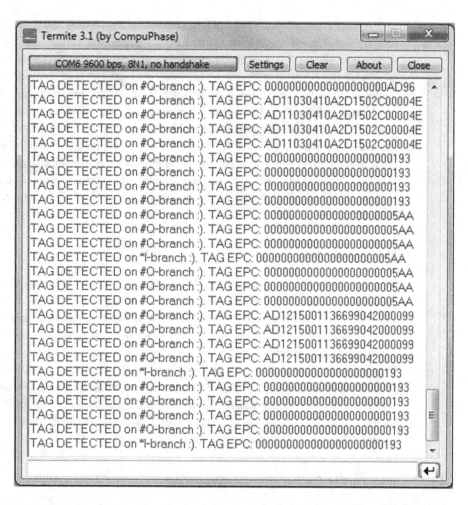

Figure 7.10. Termite RS232 terminal displaying multiple readings with indication of the reader (I or Q) branch where each transponder was singulated.

evaluate the reader performance. For the wired experiments performed in this section, the reader transmit output (output of coupler 2) was directly attached to a bare RFID test chip mounted on a PCB board via an SMA connector, and for the over-the-air experiments performed in [33], the transmit output was routed to a transmit antenna.

To adjust the power delivered to the antenna or the RFID test chip, we inserted a variable attenuator in the transmit path. The available power delivered to the antenna or the test chip was measured via a power meter connected to the coupler arm of coupler 2. In addition, we used an oscilloscope to measure the transponder baseband signal at the output of the receiver down-converter stage. In these experiments, self-jamming was suppressed by manually adjusting a load impedance tuner attached to the isolated port of the reader's directional coupler (coupler 1 in Figure 7.9) and monitoring the transmitter-to-receiver leakage on a vector signal analyzer (VSA); refer to Chapter 8 for details.

The reader communicated with a host computer via an RS232 interface. To display data from the reader (e.g., transponder EPC), we used the Termite RS232 terminal application. Figure 7.10 presents a screenshot of Termite for an over-the-air experiment with the reader in CW mode. In addition to the EPCs of transponders in the reader interrogation zone, the user interface indicates the reader branch (in-phase/quadrature) where each transponder was detected. The prevalence of quadrature branch (Q-branch) readings in the example of Figure 7.10 is because the transponder quadrature data was processed first in the algorithm of Figure 7.8. Failure to decode the quadrature data was primarily due to full dephasing between the quadrature local oscillator signal and the intercepted transponder signal, which occurs infrequently [33]. In the over-the-air experiments, this prototype achieved a communication distance of several meters.

7.4.1   Multicarrier Operation

To evaluate the RFID multicarrier backscatter scheme described in Chapter 3, we set the reader to generate multisine waveforms as described earlier in this chapter. Figure 7.11 illustrates the multicarrier generation. The bandpass nine-tone multisine at the reader transmitter output (Figure 7.11(b)) results from up-converting a four-tone baseband multisine (plus a DC component) generated by the multisine DAC (see Figure 7.11(a)) with an RF-modulated CW carrier [33]. The bandpass multicarrier was then amplified and radiated. Figure 7.11(c) compares a PIE-encoded, ASK-modulated signal using a CW carrier and multicarrier with the same average power. Notice the larger time-domain peaks (and larger PAPR) in the multicarrier case, which can improve the conversion efficiency at the transponder [14, 15].

7.4.2   Multicarrier Backscatter Demodulation

When illuminated with a multicarrier signal, a transponder will backscatter its information on all the subcarriers within its modulating bandwidth. As mentioned in Chapter 3, the first step in demodulating the transponder message backscattered on a multicarrier is to down-convert the signal using the same local oscillator that was used to generate the transmit multicarrier. Figure 7.12(a) shows the spectrum of the transponder baseband signal at the output of the in-phase down-converter mixer of our reader prototype when radiating a multicarrier waveform like the one in Figure 7.11 (c). This corresponds to an FM0-encoded transponder response at a backscatter link frequency (BLK) of 160 kHz modulated on all the illuminating subcarriers. We used an oscilloscope to digitize the transponder baseband response at 50 MSPS (see Figure 7.9) and we then processed the stored waveform offline in MATLAB.

As stated in Chapter 3, if the illuminating multicarrier is symmetrical and has a frequency component at the local oscillator frequency, then the easiest (but not optimal) way to retrieve the backscattered transponder message is to simply low-pass filter the down-converted signal with a cut-off frequency less than the subcarrier spacing $\Delta\omega$. Applying a low-pass filter to the signal in Figure 7.12(a) produced the

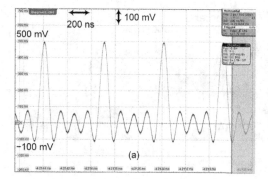

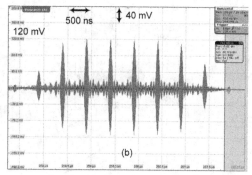

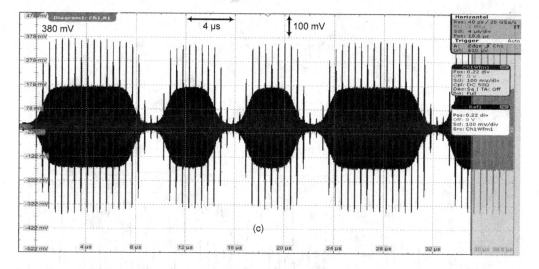

Figure 7.11 Transmit waveforms. (a) Four-tone baseband multisine at the output of the multisine DAC (DAC M in Figures 7.3 and 7.4). (b) Resulting bandpass-modulated multicarrier pulse. (c) RFID PIE-encoded, ASK-modulated symbols using a CW carrier (solid black signal) and multicarrier (pulsed signal) with the same average power.

signal in Figure 7.12(b). In the presence of noise, the transponder message on the center subcarrier can be recovered optimally by applying a matched filter with a constant amplitude template pulse, $n = 0$ in (7.2).

To retrieve the transponder information modulated on a side subcarrier ($n > 0$), we can use a matched filter whose template pulse corresponds to the subcarrier we wish to demodulate. Template pulse examples are defined as in (7.2) and illustrated in Figure 7.13. In our example, subcarriers 1, 2, 3, and 4 in Figure 7.12(a) should be filtered using the 2 MHz, 4 MHz, 6 MHz and 8 MHz pulses in Figure 7.13, respectively. To recover the information on all the subcarriers at once, we can use a baseband multisine pulse identical to the one used in the transmitter (7.3) as the template for the matched filter.

The multicarrier matched filtering could be implemented efficiently using an FFT. While FFT computation overhead is generally not a problem for offline postprocessing using, for example, MATLAB, it can be a problem for real-time applications in

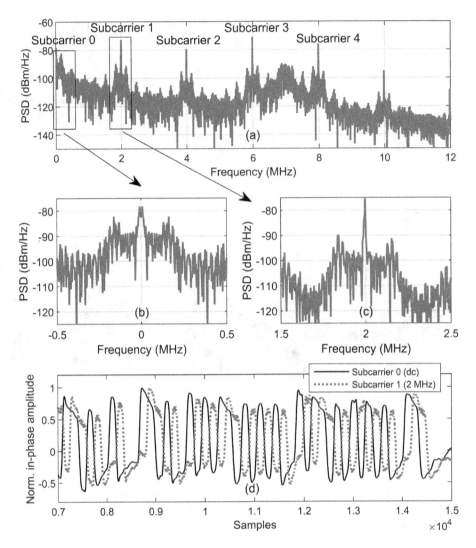

Figure 7.12 (a) Power spectral density of transponder backscatter multicarrier. (b) Magnification of transponder response on the center subcarrier (at 0 Hz). (c) Magnification of transponder response on the first side subcarrier (at 2 MHz). Note the spectrum asymmetry, which was alluded to in Chapter 3. (d) Retrieved baseband time-domain waveforms for the signals of Figures 7.11(b) and (c).

$$S_{pulse}^{n}(t) = \begin{cases} A_n \cos(2n\Delta\omega t), & 0 \leq t \leq \dfrac{T_{symbol}}{2}, \\ 0, \text{otherwise}, \end{cases} \tag{7.2}$$

$$S_{multisine}(t) = \begin{cases} \displaystyle\sum_{n=0}^{N} A_n \cos(2n\Delta\omega t), & 0 \leq t \leq \dfrac{T_{symbol}}{2}, \\ 0, \text{otherwise}, \end{cases} \tag{7.3}$$

where $n = 0, 1, 2, \ldots$, $\Delta\omega$ is the subcarrier spacing (2 MHz), N is the number of baseband subcarriers, and T_{symbol} is the transponder FM0 symbol period.

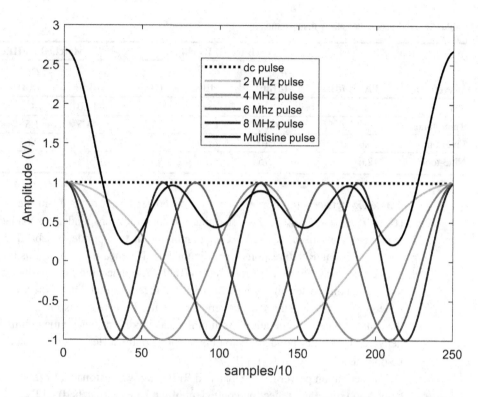

Figure 7.13 Subcarrier template pulses. Here we use a display window equal to one period of the transmit baseband multisine envelope, but for the FM0 matched filter scheme presented in Chapter 5, the duration of the template pulse should be equal to half the period of the transponder FM0 symbol, T_{symbol}:

embedded systems. Similar multicarrier techniques have been proposed for back-scatter frequency division multiple access (FDMA) for use in the internet of things (IoT) [34].

7.4.3   Transponder Sensitivity Measurements

To evaluate the transponder sensitivity enhancement provided by multicarrier signals relative to the conventional CW signal, we wired our reader to an RFID transponder chip (the G2XM RFID chip from NXP) and interrogated it with a conventional CW and several multisine waveforms. The minimum power required to activate the transponder was obtained by adjusting the transmit attenuator (Figure 7.9) – we first adjusted the attenuator so that the transponder received enough power to activate, and we then gradually increased the attenuation (decreasing the available power to the transponder) until the transponder stopped responding to the reader. At that point, we registered the available power measured by a power meter through coupler 2.

This procedure was done while the reader interrogated the transponder with a *Query* command in both CW and multisine mode. The gain/loss in transponder sensitivity of the multisine mode relative to the CW mode G_P was given by the

Table 7.4 Transponder sensitivity measurements.

		Bare RFID chip			Matched RFID chip		
Waveform	PAPR (dB)	P_{min} (dBm)	G_P (dB)	r (%)	P_{min} (dBm)	G_P (dB)	r (%)
CW	3.0	−4.45	—	—	−13.0	—	—
Three-tone	7.8	−6.0	1.55	19.5	−14.5	1.5	18.9
Five-tone	10.0	−7.10	2.65	35.7	−15.6	2.6	34.9
Nine-tone	12.6	−7.75	3.3	46.2	−16.3	3.3	46.2

difference between the minimum activation power levels obtained in the two cases $(G_P = P_{CWmin} - P_{MSmin})$. To determine whether the transponder responded to the reader query, we used the oscilloscope to monitor the transponder baseband waveform and check for an RN16 response (see Figure 7.9). The measurement results for various multicarrier waveforms are presented in Table 7.4. The nine-tone multicarrier provided a maximum sensitivity gain of 3.3 dB, suggesting that half of the power would be required to achieve the same coverage range for the tested transponder compared to the conventional CW operation. Assuming ideal conditions, where maximum PAPR is delivered to the transponder, a maximum relative multicarrier coverage range gain can be obtained as $r = \mathrm{sqrt}(G_p) - 1$ [33].

The activation power for the matched RFID chip was estimated from the measured activation power and reflection coefficient of the bare (unmatched) RFID chip [(11.3) in Chapter 11].

7.5   Conclusions

We have described the design, prototyping, and testing of a full-stack software-defined-radio-based RFID reader compliant with ISO 18000-63. This work highlights some of the key advantages of the SDR approach, including the flexibility to generate custom RFID powering waveforms, and provides a foundation for other custom designs and experiments. We used the developed reader platform to explore unconventional techniques like high-PAPR multicarrier wireless power for RFID, which can improve energy conversion efficiency in passive RFID transponders and potentially extend their coverage range. In addition to its potential for improving wireless power transfer in passive RFID systems [14, 15], the multicarrier approach enables a range of interesting new applications for backscatter medium access [34], medical devices [35, 36], wake-up radios [37], precision localization [38], and encoding and modulation schemes [39–41].

References

[1] P. Cruz, N. Borges Carvalho, and K. A. Remley, "Designing and Testing Software-Defined Radios," *IEEE Microwave Magazine*, 11(4):83–94, 2010.

[2] D. Sinha, A. K. Verma, and S. Kumar, "Software Defined Radio: Operation, Challenges and Possible Solutions," in *Tenth International Conference on Intelligent Systems and Control (ISCO)*, 2016, pp. 1–5.

[3] J. Mitola, "The Software Radio Architecture," *IEEE Communications Magazine*, 33 (5):26–38, 1995.

[4] T. F. Collins, R. Getz, D. Pu, and A. M. Wyglinski, Software-Defined Radio for Engineers. Artech House, London, 2018.

[5] A. Oliveira et al., "All-Digital RFID Readers: An RFID Reader Implemented on an FPGA Chip and/or Embedded Processor," *IEEE Microwave Magazine*, 22(3):18–24, 2021.

[6] M. Buettner and D. Wetherall, "A 'Gen 2' RFID Monitor Based on the USRP," *ACM SIGCOMM Computer Communication Review*, 40(3):41–47, 2010.

[7] N. Kargas, F. Mavromatis, and A. Bletsas, "Fully-Coherent Reader with Commodity SDR for Gen2 FM0 and Computational RFID," *IEEE Wireless Communications Letters*, 4 (6):617–620, 2015.

[8] W. Yuechun et al., "A Flexible Software Defined Radio-Based UHF RFID Reader Based on the USRP and LabView," in *2016 International SoC Design Conference (ISOCC)*, 2016, pp. 217–218.

[9] A. Povalac, "Spatial Identification Methods and Systems for RFID Tags," Ph.D. thesis, Brno University of Technology, 2012.

[10] V. Yang, E. Zhang, and A. He, "Developing a UHF RFID Reader RF Front End with an Analog Devices' Solution," Analog Devices, Inc. technical article, 2019.

[11] Ettus Research. [Online], available at: www.ettus.com/

[12] GNU Radio. [Online], available at: www.gnuradio.org/

[13] National Instruments, "LabVIEW." [Online], available at: www.ni.com/en-us/shop/labview.html

[14] M. S. Trotter and G. D. Durgin, "Survey of Range Improvement of Commercial RFID Tags with Power Optimized Waveforms," in *2010 IEEE International Conference on RFID*, 2010, pp. 195–202.

[15] A. J. Soares Boaventura and N. Borges Carvalho, "Extending Reading Range of Commercial RFID Readers," *IEEE Transactions on Microwave Theory and Techniques*, 61(1):633–640, 2013.

[16] Analog Devices, "900 MHz ISM Band Analog RF Front End." [Online], available at: www.analog.com/en/products/adf9010.html#product-overview

[17] A. J. Soares Boaventura, A. Collado, A. Georgiadis, and N. Borges Carvalho, "Spatial Power Combining of Multi-Sine Signals for Wireless Power Transmission Applications," *IEEE Transactions on Microwave Theory and Techniques*, 62(4):1022–1030, 2014.

[18] Analog Devices, "8-Bit, 125 MSPS Dual TxDAC+ Digital-to-Analog Converter." [Online], available at: www.analog.com/en/products/ad9709.html#product-overview

[19] Texas Instruments, "Dual-Channel, 12-Bit, 275-MSPS Digital-to-Analog Converter." [Online], available at: www.ti.com/product/DAC5662

[20] Texas Instruments, "DAC5662 Dual-Channel, 12-Bit, 275-MSPS Digital-to-Analog Converter (DAC) Evaluation Module." [Online], available at: www.ti.com/tool/DAC5662EVM

[21] Texas Instruments, "C2000, 32-bit MCU with 800 MIPS, 2×CPU, 2×CLA, FPU, TMU, 1024 KB flash, EMIF, 16b ADC." [Online], available at: www.ti.com/product/TMS320F28377D

[22] Texas Instruments, "F28379D Delfino Experimenter Kit." [Online], available at: www.ti.com/tool/TMDSDOCK28379D

[23] RN2 Technologies. [Online], available at: www.rfmw.com/manufacturer/rn2

[24] Skyworks, "SKY65111–348LF, 2 Watt InGaP HBT Power Amplifier ISM 800–1000 MHz Band." [Online], available at: www.skyworksinc.com/-/media/SkyWorks/ Documents/Products/101-200/200428E.pdf

[25] Analog Devices, "700 MHz to 2700 MHz Quadrature Demodulator." [Online], available at: www.analog.com/en/products/adl5382.html

[26] A. Collado and A. Georgiadis, "Improving Wireless Power Transmission Efficiency Using Chaotic Waveforms," in *International Microwave Symposium*, 2012.

[27] C.-C. Lo, et al., "Novel Wireless Impulsive Power Transmission to Enhance the Conversion Efficiency for Low Input Power," in *IEEE MTT-S International Microwave Workshop Series on Innovative Wireless Power Transmission*, 2011, pp. 55–58.

[28] M. Reynolds et al., "Multi-Band, Low-Cost EPC Tag Reader," Auto-ID Center White Paper, Massachusetts Institute of Technology, 2002.

[29] J.-H. Bae et al., "Study on the Demodulation Structure of Reader Receiver in a Passive RFID Environment." *Progress in Electromagnetics Research*, 91:243–258, 2009.

[30] C. Jin and S. H. Cho, "A Robust Baseband Demodulator for ISO 18000-6C RFID Reader Systems," *International Journal of Distributed Sensor Networks*, 8(9):406710, 2012.

[31] C. Huang and H. Min, "A New Method of Synchronization for RFID Digital Receivers," in *Eighth International Conference on Solid-State and Integrated Circuit Technology Proceedings*, 2006, pp. 1595–1597.

[32] F. Zheng and T. Kaiser, *Digital Signal Processing for RFID*, 1st ed. Wiley, Chichester, 2016.

[33] A. Soares Boaventura and N. Borges Carvalho, "The Design of a High-Performance Multisine RFID Reader," *IEEE Transactions on Microwave Theory and Techniques*, 65 (9):3389–3400, 2017.

[34] W., Liu, K., Huang, X. Zhou, and S. Durrani, "Next Generation Backscatter Communication: Systems, Techniques, and Applications." *Journal on Wireless Communications and Networking*, 2019:69, 2019.

[35] H. Zhang et al., "Wireless Power Transfer Antenna Alignment Using Intermodulation for Two-Tone Powered Implantable Medical Devices," *IEEE Transactions on Microwave Theory and Techniques*, 67(5):1708–1716, 2019.

[36] B. Wang and H. Zhang, "Ultra-Wide Dynamic Range Rectifier Topology for Multi-Sine Wireless Powered Endoscopic Capsules," in *2018 IEEE MTT-S International Wireless Symposium*, 2018, pp. 1–4.

[37] F. Hutu and G. Villemaud, "On the Use of the FBMC Modulation to Increase the Performance of a Wake-Up Radio," in *IEEE Radio and Wireless Symposium (RWS)*, 2018, pp. 139–142.

[38] N. Decarli et al., "High-Accuracy Localization of Passive Tags with Multisine Excitations," *IEEE Transactions on Microwave Theory and Techniques*, 66 (12):5894–5908, 2018.

[39] D. I. Kim, J. H. Moon, and J. J. Park, "New SWIPT Using PAPR: How it Works," *IEEE Wireless Communications Letters*, 5(6):672–675, 2016.

[40] M. Rajabi et al., "Modulation Techniques for Simultaneous Wireless Information and Power Transfer with an Integrated Rectifier–Receiver," *IEEE Transactions on Microwave Theory and Techniques*, 66(5):2373–2385, 2018.

[41] S. Claessens, N. Pan, D. Schreurs, and S. Pollin, "Multitone FSK Modulation for SWIPT," *IEEE Transactions on Microwave Theory and Techniques*, 67(5):1665–1674, 2019.

8   Self-Jamming in Backscatter Radio Systems

8.1   Introduction

In passive-backscatter RFID systems, the reader broadcasts a CW signal to wirelessly power transponders while simultaneously receiving information signals reflected by the transponders in the same frequency band. To isolate transmit and receive signals, RFID readers can use either a bistatic or a monostatic antenna scheme. The former scheme uses separate antennas for transmitting and receiving, whereas the latter shares an antenna between the transmitter and receiver through a directional coupler or circulator.

Due to imperfect isolation and clutter reflections, a large portion of the transmitted CW signal can couple into the receiver. This transmitter-to-receiver leakage or self-jamming can degrade the receiver sensitivity to the often faint signals backscattered by the transponders and is one of the main performance-limiting factors of passive-backscatter systems.

Three main mechanisms can affect reception performance in passive RFID readers. One limiting factor relates to transmitter-to-receiver leakage. Strong self-jamming can drive receivers into saturation, and this becomes more problematic if the receiver employs an RF low-noise amplifier. To be able to effectively detect weak transponder signals in the presence of strong self-jamming interference, receivers should feature a large dynamic range [1]. Another performance degradation factor relates to the direct conversion receiver architecture typically used in these systems, which can generate unwanted DC offsets at baseband in the presence of self-jamming. In this architecture, the incoming RF signal is directly converted to baseband by mixing it with a local oscillator signal at the same frequency [2]. Large self-jamming can produce DC components that are much bigger than the transponder down-converted signal at baseband. As these DC offsets are unpredictable and vary randomly over time, adaptive self-jamming cancellation is typically required. Fortunately, transponder encoding schemes such as FM0 and Miller-modulated subcarrier present minimal DC components and allow the use of high-pass filtering to remove DC offsets in direct conversion receivers.

The third major performance-limiting factor is phase noise of the reader local oscillator. The transmitted phase noise intercepted by the receiver is down-converted to baseband and can corrupt the transponder baseband response. Thus, the local oscillator phase noise is a key parameter of the RFID reader design [3].

To enhance transmitter-to-receiver isolation and improve overall system performance, high-performance passive RFID readers employ sophisticated self-jamming
suppression schemes [4–16]. Self-jamming suppression is also critical in chipless
RFID systems [17, 18], full-duplex wireless communication systems [19–21], and
radar systems [22–25]. The remainder of this chapter presents a comprehensive
discussion of self-jamming in passive-backscatter RFID systems.

8.2   Self-Jamming in Direct Conversion Receivers

In addition to the signal of interest backscattered by the transponder, RFID readers
intercept unwanted signals due to channel noise, self-jamming, clutter reflections, and
in-band interference,

$$S_{\text{received}} = S_{\text{backscatter}} + S_{\text{noise}} + \sum S_{\text{self-jamming}} + \sum S_{\text{interferer}}. \qquad (8.1)$$

As illustrated in Figure 8.1(a), self-jamming or transmitter-to-receiver leakage can be
caused by various mechanisms, including:

- crosstalk between the transmitter and receiver antennas in bistatic configurations;
- antenna mismatch and imperfect circulator isolation in monostatic configurations;
- clutter reflections from the surroundings;
- unmodulated reflections from the transponder antenna.

Moreover, the transponder-backscattered signal is typically spectrally close (tens to a few
hundred kiloHertz) to the reader carrier, which makes it susceptible to the reader local
oscillator phase noise. In-band interferers can also frequency-shift the reader local
oscillator phase noise to baseband. Figure 8.1(b) illustrates typical phase-noise leakage
mechanisms in passive-backscatter RFID systems. Self-jamming and phase-noise leakage can severely degrade the performance of RFID readers and thus proper isolation
between the transmitter and receiver is critical. In Section 8.3, we review basic transmitter-to-receiver isolation methods. In Section 8.4, we discuss more advanced techniques
for suppressing self-jamming in the RF domain. Due to changes in the RF channel, the
jamming signal is generally unpredictable. Therefore, the suppression mechanism should
be adaptive. Adaptive self-jamming suppression algorithms are studied in Section 8.5,
and strategies for mitigating DC offsets are discussed in Section 8.6.

8.3   Passive Transmitter-to-Receiver Isolation

Passive-backscatter RFID readers typically use a bistatic or monostatic antenna
scheme to isolate the transmitting and receiving paths. In bistatic architectures
(Figure 8.2(a)), self-jamming results mainly from crosstalk between the transmit and
receive antennas. In monostatic architectures (Figure 8.2(b) and (c)), self-jamming is
principally caused by antenna impedance mismatch and imperfect circulator isolation.
In both cases, clutter reflections from the surroundings can aggravate the self-jamming

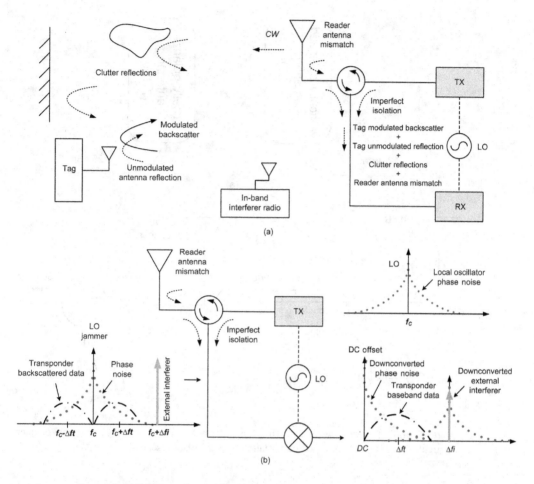

Figure 8.1 (a) Self-jamming mechanisms in a monostatic antenna configuration. (b) Illustration of local oscillator phase-noise leakage.

problem. In general, bistatic readers offer better isolation, but the monostatic imple-mentation is preferable in some applications for its reduced footprint.

A bistatic antenna experiment using two linear antennas arranged as in Figure 8.2 (a) presented an isolation of 37 dB. Figures 8.2(b) and (c) show typical parameters of two bistatic arrangements with isolations of 30 dB and 20 dB respectively for a directional coupler and circulator scheme. The jamming signal coupled into the receiver can overpower weak backscattered signals of interest (as low as −75 dBm [1] for passive transponders).

8.4   Active RF Self-Jamming Suppression

Basic isolation schemes like those above are not enough for applications requiring the highest performance. To achieve long coverage ranges and high transponder reading

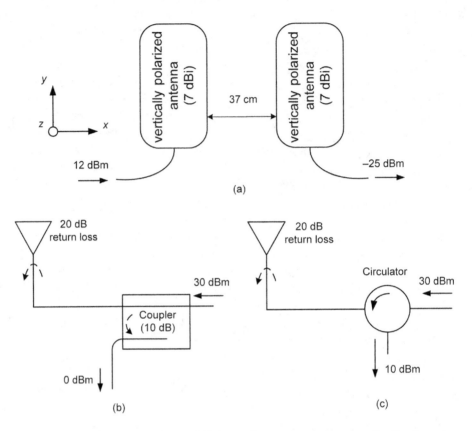

Figure 8.2 (a) Bistatic antenna scheme. (b) Monostatic antenna scheme using directional coupler as circulator. (c) Circulator-based monostatic antenna isolation scheme.

rates, the self-jamming signal component should be actively suppressed. In narrowband applications, self-jamming suppression has been shown to yield isolations of up to 80 dB [8].

Self-jamming suppression in the RF domain allows us to (i) suppress the noise associated with the jammer intercepted by the receiver, (ii) prevent saturation of the RF front end due to strong self-jamming, (iii) alleviate dynamic range requirements on the baseband circuitry and ADCs.

The basic idea behind self-jamming suppression consists of adding a signal component (the canceller signal) to the received signal to cancel the self-jamming component contained in that signal. The canceller signal is typically tapped from somewhere along the transmit chain and manipulated to get an inverted copy of the self-jamming signal. Here, we describe two common self-jamming suppression approaches used in passive-backscatter readers. In the first approach, we manipulate the canceller signal by directly adjusting its amplitude and phase (Figure 8.3), while in the second approach we do it by reflecting a replica of the transmitted signal in a controlled way (Figure 8.4).

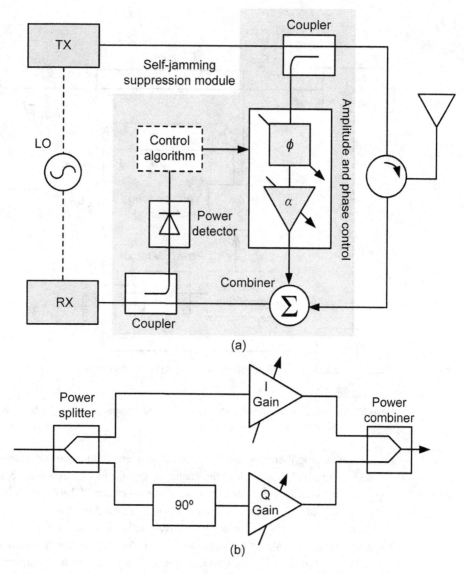

Figure 8.3 (a) Monostatic front end incorporating self-jamming suppression module based on active amplitude and phase control. The control algorithm receives its input from a power detector via an ADC and outputs control signals for the amplitude and phase of the canceller signal via DACs. See Figure 8.9 for adaptive algorithms. (b) Implementation of vector modulation for controlling the amplitude and phase of the canceller signal.

8.4.1   Classical Self-Jamming Suppression

A typical self-jamming suppression scheme based on active amplitude and phase control [4–11, 14] is depicted in Figure 8.3(a). In this scheme, the canceller signal is sampled from the transmit signal through a directional coupler and combined with the received signal using a power combiner or directional coupler. A variable phase

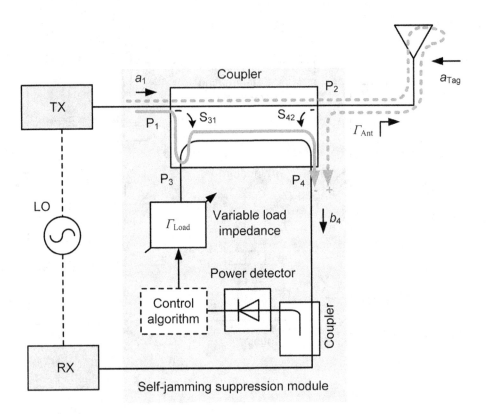

Figure 8.4 Monostatic reader front end with passive self-jamming suppression based on reflective power cancellation.

shifter and an amplifier/attenuator are used to actively control the canceller signal [9]. To achieve optimal suppression, the canceller signal must be an exact replica of the self-jamming signal except for a relative phase inversion of 180°, in which case the remaining signal will contain only the transponder signal of interest.

The combiner output is sampled by a directional coupler and RF power detector to get a received signal strength indication (RSSI). As the transponder baseband signal typically has minimal DC component, the measured RSSI is dominated by self-jamming. Based on the measured RSSI, the suppression engine acts on the canceller signal to achieve optimal suppression of the self-jammer. Alternatively, the RSSI could be obtained from the complex DC offset at the output of the receiver's quadrature down-conversion mixer. This approach does not require a dedicated power detector and enables faster self-jamming suppression algorithms [26].

To generate a canceller signal for the method presented in Figure 8.3(a), one could alternatively use a vector modulation technique like the one illustrated in Figure 8.3 (b). In this technique, a vector-modulated signal is created by splitting an incoming signal into in-phase (I) and quadrature (Q) components, which are scaled independently by variable amplifiers or attenuators and then combined to form the vector-modulated signal.

By independently scaling the I and Q branches, one can achieve arbitrary magnitude and phase of the vector-modulated signal. The Indy R2000 RFID reader from Impinj Inc. [15] features a built-in vector modulator for implementing a self-jamming suppression mechanism like that illustrated in Figure 8.3(a). Standalone vector modulators and interference cancellers are commercially available as well (e.g., QHX220, AD8340).

8.4.2   Reflective Power Cancellation

The reflective power cancellation method (illustrated in Figure 8.4) uses a directional coupler arranged as a circulator where the isolated port (P3), which is usually matched to 50 Ω, is terminated with a variable load impedance [12, 13, 16]. This allows a copy of the transmitted signal to be reflected and combined with the received signal at port P4.

The optimal canceller signal needed to suppress the self-jamming at Port P4 is achieved by adjusting the load impedance presented to port P3. Assuming[1] that the self-jamming is solely due to antenna mismatch, the incoming wave at the receiving port P4 can be written as in (8.2). Considering a symmetrical, reciprocal coupler ($S_{21} = S_{12} = S_{34} = S_{43} = T$, $S_{42} = S_{31} = C$, and $S_{41} = S_{32} = I$), then (8.2) simplifies to (8.3). If the return loss presented to the isolated port is made symmetrical to that of the antenna ($\Gamma_{\text{Load}} = -\Gamma_{\text{Ant}}$) then the leakage signal is cancelled out, and the remaining incoming wave at port P4 will consist of a scaled version of the transponder response (8.4):

$$b_4 = a_{\text{Tag}}S_{42} + a_1(S_{21}\Gamma_{\text{Ant}}S_{42} + S_{31}\Gamma_{\text{Load}}S_{43}), \tag{8.2}$$

$$b_4 = a_{\text{Tag}}S_{42} + a_1 TC(\Gamma_{\text{Ant}} + \Gamma_{\text{Load}}), \tag{8.3}$$

$$b_4 = a_{\text{Tag}}S_{42}. \tag{8.4}$$

Besides isolating the transmitting and receiving signal paths, the directional coupler is used to sample the transmit signal and combine the canceller and received signals. Compared to the previous method, reflective power cancellation can offer improved linearity and noise performance at lower complexity. To evaluate this technique, we assembled a self-jamming suppression module consisting of a custom directional coupler attached to a load impedance tuner terminated on a 50 Ω load (see Figure 8.5). This module was integrated with a custom ISO 18000-63-compliant UHF RFID reader similar to the one described in Chapter 6. The directional coupler was designed on FR4 for the 900 MHz band and presented a transmission loss, coupling, and directivity of 0.5 dB, 12 dB, and 20 dB at 860 MHz, respectively.

The coupler isolation with and without active carrier suppression was characterized using a vector network analyzer (VNA). For this experiment, an antenna with a return

---

[1] Imperfect coupler directivity and clutter reflections also contribute to self-jamming.

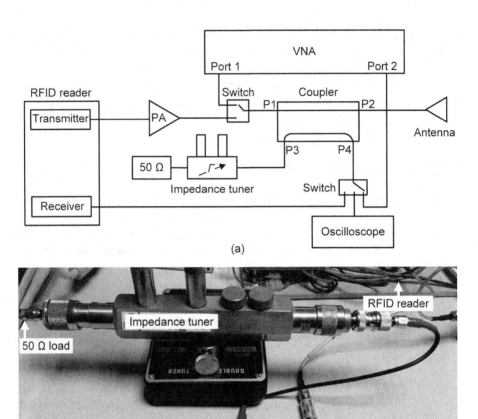

(a)

(b)

Figure 8.5 Experimental setup used to evaluate the reflective power cancellation method. (a) Block diagram of the measurement setup. (b) Photograph of the self-jamming suppression setup consisting of a directional coupler and impedance tuner terminated on a 50 Ω load.

loss better than 14 dB in the 900 MHz band was connected to the coupler port 2 and the coupler ports 1 and 4 were measured with the VNA. The coupler's isolated port was first terminated with a 50 Ω load (no active suppression) and then with the impedance tuner. In the first case, the overall isolation is set by the coupler directivity and coupling factor, and the antenna mismatch. For the second case, we achieved an isolation of up to 50 dB at 860 MHz for an optimal tuner impedance of $19.6 + 13.5j$ (see Figure 8.6).

We then used our custom RFID reader to evaluate the carrier suppression with a power of about 31 dBm at the reader PA output. The signal coupled into the receiving port P4 was measured by using both an oscilloscope and a power meter. Experimental results (see Table 8.1) showed an isolation of about 40 dB. Due to changes in the surrounding environment, the optimal impedance presented to the coupler isolating port differs from the experiment previously performed using the VNA. The load

Table 8.1 Evaluation of carrier suppression scheme using a custom RFID reader.

Tuner impedance Z_{Load} at port 3 (Ω)	P_{IN} at port 1 (dBm)	P_{CPL} at port 4 (dBm)	$ISO = S_{41} = P_{IN} - P_{CPL}$ (dB)
50	30.8	11.0	20.8
24+14j	30.8	−10.0	40.8

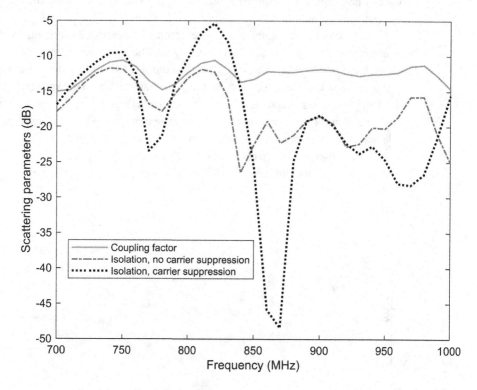

Figure 8.6 Experimental evaluation of carrier suppression. The isolation at 860 MHz was improved by about 25 dB compared to no active carrier suppression. The isolation could be further optimized by doing a fine load impedance adjustment.

impedance tuner was used in this experiment for evaluation purposes only. In real applications, the termination impedance can be implemented, for example, using PIN diodes or digitally tunable capacitors [13, 16] and controlled electronically, rather than manually.

8.5 Adaptive Self-Jamming Suppression

In the previous example we considered a static scenario, but in practice self-jamming interference varies randomly over time. Therefore, the suppression mechanism should be able to dynamically track changes and automatically update the cancellation parameters. Here, we discuss the most common adaptive algorithms used in passive RFID readers including off-the-shelf ASIC readers [15], namely full search and gradient descent search algorithms.

For adaptive self-jamming suppression, the residual signal after self-jamming cancellation (the output of the combiner in Figure 8.3(a)) is monitored to determine the optimal cancellation parameters that minimize this signal. In this approach, two control parameters are required to independently control the amplitude and phase of the canceller signal.

In the example of Figure 8.3(a), the control parameters correspond to the gain/attenuation and phase shift of the canceller signal. These parameters are directly accessed through the amplifier/attenuator and phase-shifter [9, 14]. In Figure 8.3(b), the control parameters are accessed via the in-phase and quadrature gains of the vector modulator [7, 10, 19]. In the reflective power cancellation approach (Figure 8.4), the control parameters are the real and imaginary parts of the impedance presented to the isolated port of the coupler and can be manipulated via electronically tunable capacitors [13].

For the subsequent discussion, consider the example of Figure 8.3(a). Figure 8.7(a) and (b) present the theoretical RSSI at the output of the combiner after carrier

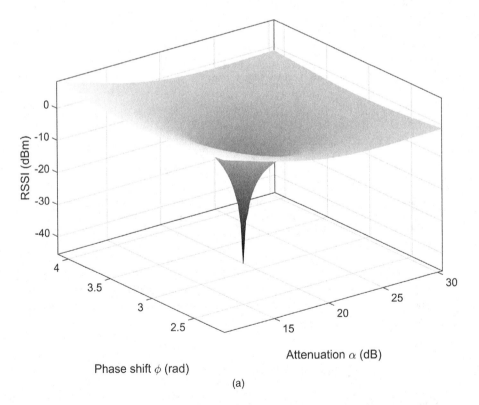

(a)

Figure 8.7 (a) Theoretical residual RSSI of received signal after suppression. (b) Heatmap of the plot in (a). The optimal phase-shift and attenuation for this example are π rad and 20 dB, respectively. (c) Theoretical suppression as a function of amplitude and phase mismatch.

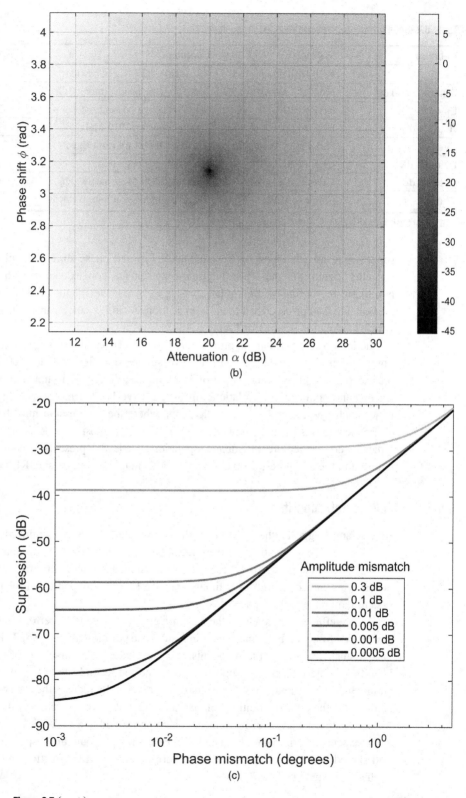

Figure 8.7 (cont.)

Table 8.2 Self-jamming suppression reported in the literature.

Isolation	System	Comment	Ref.
40 dB	UHF RFID	Broadband, 82 MHz bandwidth	[10]
Better than 30 dB	FMCW RADAR	26 GHz, wide bandwidth (1.7 GHz)	[22]
Up to 50 dB	UHF RFID	Static reflective power cancellation	This work
70 to 80 dB	UHF RFID	Single frequency, 919 MHz	[8]
45 dB	Full-duplex Wi-Fi	40 MHz OFDM transmit signal	[19]
Up to 73 dB	UHF RFID	Multi-antenna operation	[11]
50 to 65 dB	UHF RFID	Broadband, 20 MHz bandwidth	[9]
27 dB	Chipless RFID	Relative suppression improvement	[17]

OFDM – Orthogonal frequency-division multiplexing

suppression, as a function of the attenuation (α) and phase shift (ϕ) applied to the canceller signal. As the RSSI landscape presents a convex shape with a global minimum, the optimal setpoint $\left(\alpha_{\text{optimal}}, \phi_{\text{optimal}} \right)$ can be found via a minimization method like the one described in the next section. A similar curve can be drawn for the vector modulation scheme of Figure 8.4 [19].

The impact of phase and amplitude mismatch in carrier suppression is evaluated in Figure 8.7(b). As can be seen, to achieve suppression figures below 60 dB, very precise phase and amplitude control is required. While this is not a problem for narrowband applications (isolations up 80 dB have been demonstrated [8]), it is a major challenge for broadband applications, where the suppression must happen for all frequencies inside the bandwidth [27]. Fortunately, most UHF RFID applications typically do not require broadband operation. Table 8.2 presents isolation figures reported in the literature for several applications ranging from passive RFID to radar.

8.5.1   Full Search Algorithm

As its name suggests, the full search algorithm evaluates the transmitter-to-receiver isolation for all combinations of the control parameters to select the optimal setpoint. For large search spaces, this approach becomes slow and may fail to track changes in the medium and converge in a timely manner [28]. An N-point control parameter system requires a search space of N^2 points.

A more efficient approach to the search algorithm consists of performing a coarse grid search followed by a fine search in a sub-space of the original grid. This leads to only a small fraction of the N^2 points being evaluated. The first step of the search identifies the best coarse (sub-optimal) point and the second step refines the solution around that sub-optimal point. This approach dramatically reduces the time required to converge to the optimal solution compared to the brute force search over the full $N \times N$ grid. Additional levels of refinement can be performed to achieve better resolutions at the expense of computation time. To avoid overlapping and blind spots, the search grid should be carefully planned. This algorithm can be very effective and has been applied in commercial ASICs [15].

Examples of constellations for the full search algorithm based on amplitude and phase control and in-phase and quadrature control are shown in Figure 8.8. An optimal search based on a coarse search followed by a fine search like that used in [15] is illustrated in Figure 8.8(b), where the in-phase and quadrature gains of a vector modulator are used to control the canceller signal. In this example, a 7×7 coarse grid (dark dots) is first searched for a sub-optimal point, followed by a refined search in a 5×5 grid (gray dots) around the sub-optimal point. This reduces the search space from 961 down to only 74 points.

8.5.2   Gradient Descent Algorithm

The gradient descent search algorithm iteratively approximates the minimum of a function based on its gradient [7, 20]. In Figure 8.9(a), we propose an implementation of this algorithm that can be used in conjunction with the self-jamming suppression scheme previously presented in Figure 8.3(a). This algorithm starts by computing the local gradient of the RSSI function to determine the decreasing direction of the function. Based on this information, the algorithm determines the values of the control parameters for the next iteration.

To compute the local gradient, three points are sampled for RSSI. First, the current point is sampled, (α, ϕ). Then, by separately changing the two control parameters (α and ϕ) by small amounts (δ_α and δ_ϕ), two additional points are sampled: $(\alpha + \delta_\alpha, \phi)$ and $(\alpha, \phi + \delta_\phi)$. The gradient of the RSSI curve with respect to the control parameters, $[\nabla_\alpha, \nabla_\phi]$, is given by (8.5), and the values of the control parameters for the next iteration of the algorithm are given by the gradient descent equation (8.6). The control parameters are updated in the descendant direction of the RSSI function:

$$\nabla \text{RSSI}(\alpha, \phi) = \left(\nabla_\alpha, \nabla_\phi\right) = \left[\frac{\partial \text{RSSI}(\alpha, \phi)}{\partial \alpha}, \frac{\partial \text{RSSI}(\alpha, \phi)}{\partial \phi}\right]$$

$$\approx \left[\frac{\text{RSSI}(\alpha + \delta_\alpha, \phi) - \text{RSSI}(\alpha, \phi)}{\delta_\alpha}, \frac{\text{RSSI}(\alpha, \phi + \delta_\phi) - \text{RSSI}(\alpha, \phi)}{\delta_\phi}\right],$$

$$(8.5)$$

$$\begin{cases} \alpha = \alpha - \mu_\alpha \nabla_\alpha, \\ \phi = \phi - \mu_\phi \nabla_\phi, \end{cases} \qquad (8.6)$$

where the step size parameters μ_α and μ_ϕ are positive real numbers. After updating the control parameters according to (8.5) and (8.6), the new RSSI value RSSI(α, ϕ) is sampled. If it is less than the current RSSI value, the process is repeated. If the algorithm converges, then each new sampled RSSI gets closer and closer to the minimum of the RSSI function. If the new RSSI is greater than the previous RSSI, it means that the algorithm is approaching the minimum. In that case, the algorithm decreases its step sizes (both the derivation step sizes δ_α and δ_ϕ and the algorithm step sizes μ_α and μ_ϕ), reverses the direction, and continues attempting to converge to the

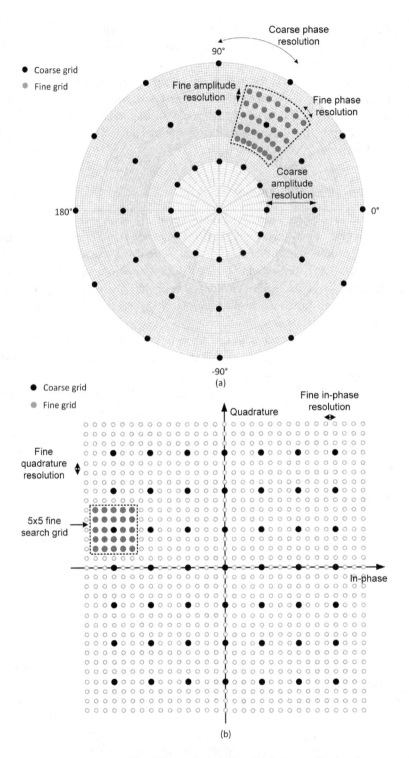

Figure 8.8 Illustration of the full search algorithm. (a) Polar amplitude–phase constellation. (b) In-phase and quadrature constellation (similar to [15]). The gray circles correspond to the fine search grid and the black dots correspond to the coarse search grid.

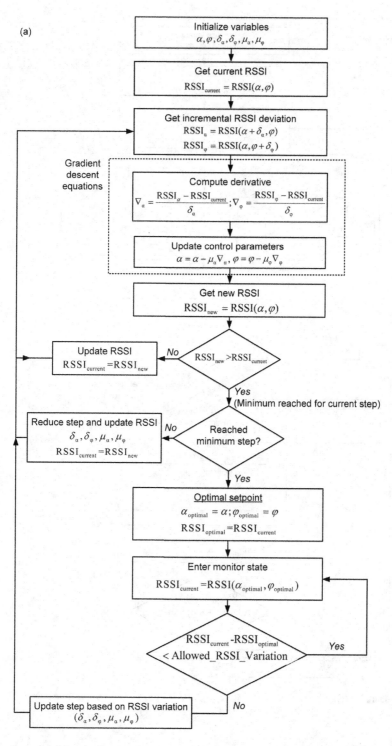

Figure 8.9 (a) Adaptive gradient descent algorithm. (b) Alternative implementation.

(b)

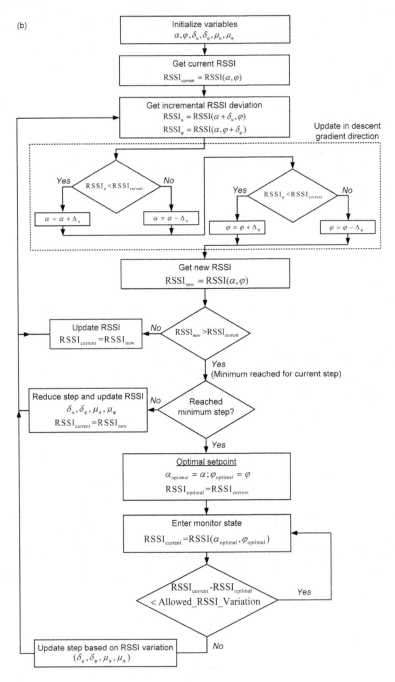

Figure 8.9 (cont.)

optimal setpoint. The optimal setpoint is found when the algorithm reaches its minimum step size.

Using variable step sizes improves the algorithm's performance. The step size is initialized with a relatively large value and gradually decreased as we approach the minimum. Once a global minimum is found, the optimal control parameters remain constant and the RSSI is continually monitored and compared to the optimal point. If a change is detected and the difference exceeds a predefined threshold, the step size is updated based on the difference and the search is restarted. Factoring in the deviation from the optimal setpoint significantly improves the algorithm's performance – small deviations do not require a large step size that would otherwise increase the convergence time unnecessarily. To prevent false alarms caused by noisy local minima, minimum values above a certain threshold are ignored.

An alternative implementation of the algorithm just described is presented in Figure 8.9(b). Note that these algorithms can also be applied to self-jamming suppression based on vector modulation or reflective power cancellation. For further reading on adaptive self-jamming suppression algorithms, refer to [26, 28].

8.6   Baseband Offset Removal

If no RF self-jamming suppression is employed, large DC offsets can appear at the baseband circuitry, posing stringent dynamic range requirements on the baseband circuitry, including amplifiers and ADCs. For systems with enough dynamic range, DC offsets can be handled in the digital domain. In [29], we used a finite impulse response (FIR) filter implemented in an FPGA to eliminate DC offsets in the digital domain. But DC offset removal in the analog domain can loosen the dynamic range requirements of the baseband electronics and is preferable for designs with limited baseband processing resources.

As mentioned before, transponder encoding schemes like FM0 or Miller-modulated subcarrier code present minimal DC components, which allows the use of high-pass filtering to suppress DC offsets in direct conversion receivers. But simple DC blocking filters may introduce transients that can distort the transponder baseband response. This can be mitigated by using switched non-linear high-pass filtering (described next), which allows the removal of DC offsets while compensating for transients.

8.6.1   Baseband Transients

DC blocking filters are typically used at the output of down-conversion mixers in direct conversion receivers to suppress DC offsets. The filter capacitor is

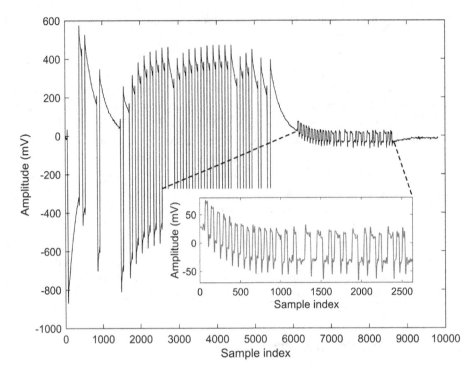

Figure 8.10 Down-converted reader command followed by transponder response at the output of the DC blocking filter. The inset shows a close-up of the distorted transponder response.

designed to account for the time constant and attenuation of the filter and transponder data rate. For the data rates used in our experiments (40 kbps to 640 kbps), a 100 nF DC blocking capacitor was used. To show the distortion introduced by the DC blocking filter in the signal of interest, we interrogated an ISO 18000-63 transponder using a custom RFID reader and examined the transponder baseband response waveform. The transponder was commanded to reply with FM0 encoding at 320 kbps and to prepend a pilot tone to its message. Figure 8.10 shows the signal at the output of the DC blocking filter, where the distortion introduced by the filter is evident in the inset. Small to moderate distortion can be mitigated by discarding the pilot data and processing only the transponder payload data. Larger distortions require more processing as discussed in the next sections and in [30].

8.6.2   Digital Transient Compensation

A simple approach to compensate for the transient response of the DC blocking filter in the digital domain is to subtract the moving average of the transient-distorted signal from the signal,

$$S_{\text{Corrected}}(n) = S_{\text{Baseband}}(n) - \frac{1}{\omega}\left\{S_{\text{Baseband}}\left(n - \frac{\omega - 1}{2}\right) \cdots + S_{\text{Baseband}}(n - 1)\right.$$

$$\left. + S_{\text{Baseband}}(n) + S_{\text{Baseband}}(n + 1) \cdots + S_{\text{Baseband}}\left(n + \frac{\omega - 1}{2}\right)\right\},$$

$$(8.7)$$

where n is the sample index, the subtracting term is the moving average of the transient-distorted signal, and w is the window of the average, which determines the cut-off frequency of the low-pass filtering operation performed by this filter. Increasing the filter window provides a smoother average curve but increases the computation time. This filter requires that at least $w/2$ samples be stored in memory.

We used MATLAB to post-process transponder baseband data collected with our custom RFID reader and apply transient correction. Figure 8.11(a) shows the transponder response signal that results from applying the correction in (8.7) to the distorted signal in Figure 8.10, and Figure 8.11(b) shows typical transponder constellations before and after transient correction. After correction, the constellation is recentered about the origin.

8.6.3   Non-Linear Filtering

A switched high-pass filter can be used to compensate for the DC blocking filter transient in the analog domain by pre-charging the capacitor to the correct value during the reader CW transmission. For that, we use two NMOS transistor switches, which are driven by the reader's digital signal processor, to control the isolation and initial state of the filter. The operation of this filter is illustrated in Figure 8.12.

A series and a shunt switch (S1 and S2) are used to isolate the receiver during the reader transmission and control the charging of the DC blocking capacitor. During the reader transmission, S1 is open and S2 is closed to isolate the receiver baseband circuit. Immediately after the reader transmission, both S1 and S2 are closed so that the coupling capacitor is grounded and pre-charged with the DC offset generated by the reader CW leakage.

During the transponder response, S1 remains closed and S2 is open, allowing the pre-charged DC offset to be subtracted from the incoming signal at the beginning of the transponder response. This removes the transient and centers the output signal. The RF front end [31] used in Chapter 7 employs a similar mechanism.

8.7   Conclusions

This chapter provided a comprehensive discussion on self-jamming in passive-backscatter systems. We described common self-jamming and interference mechanisms, transmitter-to-receiver isolation approaches, RF self-jamming suppression, DC offset removal, and baseband transient compensation. We also discussed techniques used in commercial ASIC RFID readers and experimentally demonstrated self-jamming suppression up to 50 dB.

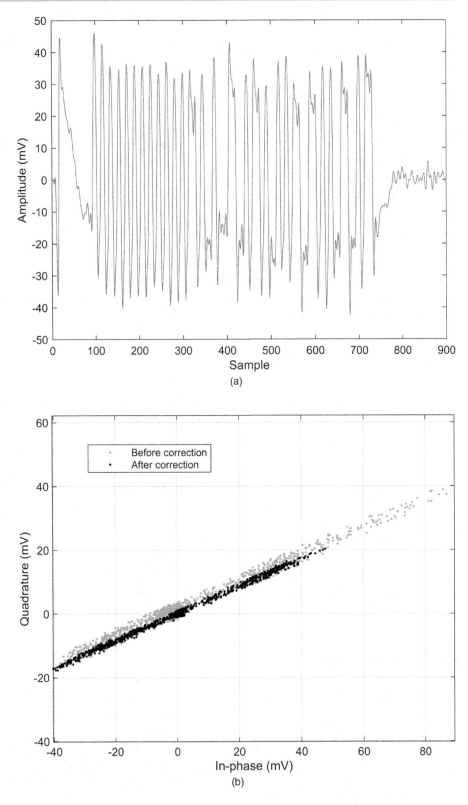

Figure 8.11 (a) Transient corrected waveform. (b) Constellations before and after transient correction.

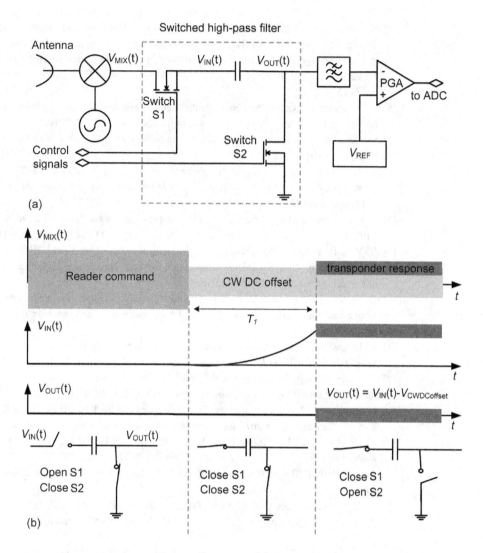

Figure 8.12 (a) Baseband stage incorporating switched high-pass filter. (b) Illustration of switched high-pass filter operation.

References

[1] J. Wang, et al., "System Design Considerations of Highly-Integrated UHF RFID Reader Transceiver RF Front-End," in *Ninth International Conference on Solid-State and Integrated-Circuit Technology*, 2008, pp. 1560–1563.

[2] P. Cruz, H. Gomes, and N. Carvalho, "Receiver Front-End Architectures – Analysis and Evaluation," in *Advanced Microwave and Millimeter Wave Technologies: Semiconductor Devices Circuits and Systems*, ed. M. Mukherjee. Intech, Rijeka, Croatia, 2012, pp. 495–520.

[3] N. Usachev et al., "System Design Considerations of Universal UHF RFID Reader Transceiver ICs," *Facta Universitatis, Electronics and Energetics*, 28(2):297–307, 2015.

[4] R. Langwieser, et al., "A Modular UHF Reader Frontend for a Flexible RFID Testbed," in *Proceedings of the International EURASIP Workshop on RFID Technologies*, 2008, pp. 1–12.

[5] J. Posamentier, "Radio Frequency Identification Apparatus, System and Method Adapted for Self-Jammer Cancellation," U.S. patent number 7684751 B2, 2010.

[6] S. Chiu, M. Sajid, and I. Kipnis, "Cancelling Self-Jammer Signals in an RFID System," U.S. patent 8013715 B2, 2011.

[7] I. Mayordomo and J. Bernhard, "Implementation of an Adaptive Leakage Cancellation Control for Passive UHF RFID Readers," in *IEEE International Conference on RFID*, 2011, pp. 121–127.

[8] D. P. Villame and J. S. Marciano, "Carrier Suppression Locked Loop Mechanism for UHF RFID Readers," in *IEEE International Conference on RFID 2010*, 2010, pp. 141–145.

[9] B. You, B. Yang, X. Wen, and L. Qu, "Implementation of Low-Cost UHF RFID Reader Front-Ends with Carrier Leakage Suppression Circuit," *International Journal of Antennas and Propagation*, 2013:135203, 2013.

[10] G. Lasser, R. Langwieser, and A. L. Scholtz, "Broadband Suppression Properties of Active Leaking Carrier Cancellers," in *IEEE International Conference on RFID*, 2009, pp. 208–212.

[11] R. Langwieser et al., "Active Carrier Compensation for a Multi-Antenna RFID Reader Frontend," in *IEEE MTT-S International Microwave Symposium*, 2010, pp. 1532–1535.

[12] T. Brauner and X. Zhao, "A Novel Carrier Suppression Method for RFID," *IEEE Microwave and Wireless Components Letters*, 19(3):128–130, 2009.

[13] M. Koller and R. Küng, "Adaptive Carrier Suppression for UHF RFID Using Digitally Tunable Capacitors," in *2013 European Microwave Conference*, 2013, pp. 943–946.

[14] J. Lee et al., "A UHF Mobile RFID Reader IC with Self-Leakage Canceller," in *IEEE Radio Frequency Integrated Circuits Symposium*, 2007, pp. 273–276.

[15] Impinj, "Indy R2000 UHF Gen 2 RFID Reader Chip." [Online], available at: www.impinj .com

[16] E. A. Keehr, "A Low-Cost Software-Defined UHF RFID Reader with Active Transmit Leakage Cancellation," in *2018 IEEE International Conference on RFID*, 2018, pp. 1–8.

[17] R. R. Antayhua, C. R. Rambo, and F. R. de Sousa, "Self-Interference Cancellation in Chipless RFID Readers for Reading Range Enhancement," in *IEEE International Instrumentation and Measurement Technology Conference*, 2018, pp. 1–6.1

[18] M. Forouzandeh and N. Karmakar, "Self-Interference Cancelation in Frequency-Domain Chipless RFID Readers," *IEEE Transactions on Microwave Theory and Techniques*, 67 (5):1994–2009, 2019.

[19] M. Jain et. al., "Practical, Real-Time, Full Duplex Wireless," in *Proc. MobiCom '11*, 2011, 301–312,

[20] M. Jain, "Single Channel Full-Duplex Radios." Ph.D. thesis, Stanford University, 2011.

[21] D. Bharadia, E. McMilin, and S. Katti, "Full Duplex Radios," *ACM SIGCOMM Computer Communication Review*, 43(4):375–386, 2013.

[22] K. Lin, Y. E. Wang, C.-K. Pao, and Y.-C. Shih, "A Ka-Band FMCW Radar Front-End With Adaptive Leakage Cancellation," *IEEE Transactions on Microwave Theory and Techniques*, 54(12):4041–4048, 2006.

[23] C. Will et. al., "A 24 GHz Waveguide Based Radar System Using an Advanced Algorithm for I/Q Offset Cancelation," *Advances in Radio Science*, 15:249–258, 2017.

[24] S. Mann et. al., "High-Precision Interferometric Radar for Sheet Thickness Monitoring," *IEEE Transactions on Microwave Theory and Techniques*, 66(6):3153–3166, 2018.

[25] M. Pehlivan and K. Yegin, "Self-Jamming and Interference Cancellation Techniques for Continuous Wave Bi-static Radar Systems," in *22nd International Microwave and Radar Conference (MIKON)*, 2018, pp. 255–257.

[26] G. Lasser, R. Langwieser, and C. F. Mecklenbräuker, "Automatic Leaking Carrier Canceller Adjustment Techniques," *EURASIP Journal on Embedded Systems*, 2013:8, 2013.

[27] Y. Deng, Active Interference Cancellation, Technologies Comparisons. White paper, Bascom Hunter, 2015.

[28] G. Lasser et. al., "Fast Algorithm for Leaking Carrier Canceller Adjustment," in *Fourth International EURASIP Workshop on RFID Technology*, 2012, pp. 46–51.

[29] A. Soares Boaventura et al., "Perfect Isolation: Dealing with Self-Jamming in Passive RFID Systems," *IEEE Microwave Magazine*, 17(11):20–39, 2016.

[30] J.-H. Bae et al., "Study on the Demodulation Structure of Reader Receiver in a Passive RFID Environment." *Progress in Electromagnetics Research*, 91:243–258, 2009.

[31] Analog Devices, "900 MHz ISM Band Analog RF Front End." [Online], available at: www.analog.com/en/products/adf9010.html#product-overview

9   Wake-Up Radios for IoT Applications

9.1   Introduction

Conventional radios typically use power-hungry transceivers, which makes them not ideal for battery-powered low-maintenance IoT applications. To extend the battery lifespan in these applications, medium access control (MAC) schemes [1] and energy scavenging [2] have been investigated. In duty cycle MAC protocols, commonly used in wireless sensor networks (WSNs), the radio transceiver is periodically turned on and off to conserve energy [1]. The more the radio stays off, the more energy is saved, but the higher the communication latency. Low energy consumption and low communication latency are therefore conflicting requirements in duty cycle MAC protocols.

To simultaneously reduce power consumption and latency in low data traffic applications, the concept of wake-up radio has been proposed where a companion low-power, low-data-rate wake-up radio receiver is used to asynchronously wake up the main radio only when necessary (see Figure 9.1). Although the wake-up radio concept has been fairly well investigated in the past decade [3–13], it has recently gained renewed interest in the context of the IoT [14, 15] and the Institute of Electrical and Electronics Engineers (IEEE) has recently begun efforts on the standardization of wake-up receivers for Wi-Fi devices [15, 16].

In this chapter, we present a discussion of wake-up radios, including a literature review, and we give a glance at the new IEEE Wi-Fi wake-up radio standard. We then describe the implementation of a custom wake-up radio receiver with addressing capabilities operating in the 900 MHz ISM band, and we discuss its integration into a wireless sensor network system.

9.2   Literature Review

Most commercially available wake-up radio solutions operate in the 125 kHz ISM band and are typically devoted to short- to medium-range passive keyless entry and access control applications. Examples of fully integrated solutions in the 125 kHz band include the AS3931 IC [3] (used in [7]), ATA5282 IC, MCP2030 IC [4], and EM4083 IC [5]. Some solutions provide baseband capabilities to be integrated with external RF circuitry [6].

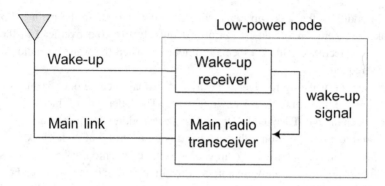

Figure 9.1 Wake-up radio concept.

Even though some RFID manufacturers have recently incorporated wake-up radio features in UHF transponders [8], most of the work on wake-up radio in the 900 MHz and 2.4 GHz bands has been limited to research and academia [9–12]. Proposed wake-up radio concepts in these bands feature addressing capabilities, communication distances of several meters, power consumption as low as few microwatts and fast wake-up times.

9.2.1 The IEEE 802.11ba Standard for IoT Wi-Fi

IEEE 802.11 Wi-Fi is a compelling choice for many IoT application segments for various reasons: it is a widely adopted standard, 802.11-compliant devices can be seamlessly integrated with the IEEE 802 ecosystem (including Ethernet), and most of the required infrastructure is already deployed. However, being originally designed for high-speed communications in personal computers and portable devices where energy consumption is typically not a major concern, the conventional Wi-Fi standard overlooks the energy consumption aspect. To remedy this shortcoming and make Wi-Fi a stronger candidate for low-power low-maintenance IoT applications, IEEE established the 802.11ba Wake-Up Radio Task Group in 2016 to develop the low-power wake-up radio (LP-WUR) standard for Wi-Fi networks [15, 16].

9.3 Wake-Up Radio Design Requirements

Power consumption is the most important requirement of a wake-up radio receiver. To be effective, the wake-up radio must consume several orders of magnitude less than the main radio. For example, the IEEE 802.11ba standard recommends a wake-up radio consumption of less than 100 µW in the active state, which contrasts with the consumption of legacy Wi-Fi devices in the range of tens to hundreds of milliwatts [17].

In addition, the wake-up range should be comparable to the main radio, and the wake-up operation should be performed quickly to avoid overhead in the primary radio connectivity. This is especially important when the wake-up radio is activated very frequently.

Since the wake-up receiver is always on, it may receive out-of-band and in-band interference, including from its own network. The latter type of interference poses the biggest challenge. To avoid false in-band alarms, addressing mechanisms are typically required to ensure that only the addressed node or group of nodes is woken up.

Finally, the size, cost, and complexity of the companion wake-up radio should be as low as possible to allow seamless integration into existing systems. For instance, the IEEE 802.11ba standard specifies a simple wake-up protocol and signaling based on on–off keying modulation and envelope detection.

9.4   A Wake-Up Radio Protocol

We have developed a custom wake-up radio system operating in the 900 MHz ISM band for use in a wireless sensor network system. The proposed wake-up radio architecture is depicted in Figure 9.2. The companion wake-up radio comprises an antenna, impedance-matching circuit, envelope detector, low-power comparator (MAX9021 [18]), and microcontroller unit (MCU; ATmega16 [19]). In addition to boosting the energy transference between the antenna and the envelope detector, the impedance-matching circuit provides frequency selectivity that helps to suppress out-of-band interference. Therefore, a high-quality factor bandpass characteristic is desirable for the impedance-matching network.

We use an envelope detector circuit based on a two-stage (four-diode) Dickson charge pump using HSMS2852 diode pairs [20] to demodulate the incoming OOK-

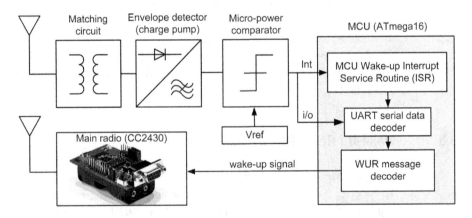

Figure 9.2 Block diagram of the proposed wake-up radio system including the main radio transceiver (CC2430 from Texas Instruments). The envelope detector demodulator uses a two-stage charge pump circuit.

modulated RF signal and extract the wake-up baseband envelope. The envelope detector output is then digitized by the comparator and fed into the MCU through an interrupt line (for waking up the MCU) and an I/O line (for processing the wake-up baseband data).

The minimum RF power required to wake up the MCU depends primarily on the sensitivity of the envelope detector circuit, and in our prototype the threshold level of the comparator circuit is adjusted via a potentiometer. Improved performance can be achieved by using a data-slicing demodulation technique like the one described in Chapter 6. After waking up, the MCU implements the wake-up radio protocol (described next).

9.4.1   Wake-Up Protocol

The proposed wake-up radio protocol is illustrated in Figure 9.3. By default, the wake-up radio receiver is in standby mode with minimal power consumption, the main radio

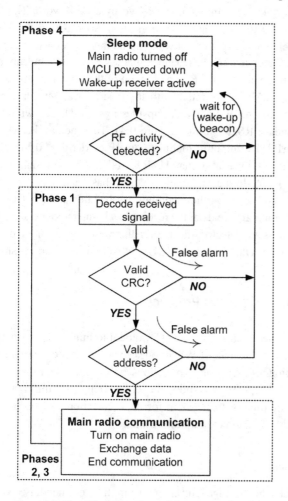

Figure 9.3 Simplified flowchart of the proposed wake-up radio protocol.

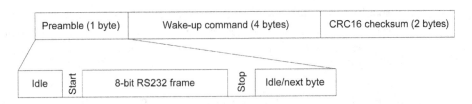

Figure 9.4 Wake-up radio message. Each message byte is encapsulated in a standard RS232 frame.

transceiver is turned off, and the MCU is in a low-energy power-down mode from which it can be woken via an external interrupt. Upon detecting in-band RF power, the wake-up receiver generates a positive transition that wakes up the MCU. The MCU then processes the subsequent incoming signal received through the I/O line to detect the wake-up message. The wake-up message (Figure 9.4) comprises a one-byte preamble, a four-byte wake-up command, and a two-byte CRC word. The preamble (a 0xFF code) is intended to create an RF carrier signal to wake up the MCU. The wake-up command field can carry either a wake-up address or a custom command.

To encode/decode the wake-up baseband signal, we use the bult-in universal asynchronous receiver/transmitter (UART) peripheral of the MCU. On the transmitter side, we configure the UART transmitter with the desired wake-up data rate (e.g., 9.6 kbps) and transmit the seven-byte wake-up message via UART with each byte encapsulated in a standard RS232 frame (Figure 9.4). To generate the RF wake-up signal, we use the UART Tx output to modulate an RF carrier at 860 MHz.

On the receiver side, we use a similar UART, set at the same data rate as the transmitting UART, to decode the incoming wake-up baseband signal connected to the UART Rx input. After decoding the incoming baseband data, the MCU processes the CRC and wake-up command fields of the received wake-up radio message. If a valid CRC and wake-up address are detected, the main radio transceiver is turned on for high-speed data communication. Otherwise, the MCU returns to the low-power sleep mode.

9.5   Classical MAC versus Wake-up Protocol

In [21], the authors applied a duty cycle MAC protocol to minimize power consumption and extend the battery lifespan of WSN nodes for wireless localization. Here, we consider the integration of our wake-up radio scheme into the radio transceiver CC2430 [22] used in [21, 23]. To show the effectiveness of this approach, we estimated the battery lifetime of that system for the conventional MAC protocol and our wake-up radio scheme, and compare the results.

9.5.1   MAC Battery Lifespan

Equation (9.1) models the average current consumption of the duty cycle MAC protocol used in [22]:

Table 9.1 Consumption parameters of the duty cycle MAC protocol used in [22].

Description	Consumption (mA)	Time interval (ms)
MCU active	12.5	4.16
Radio in Rx mode	31	1.6
Radio in Tx mode	29	1.12
Sleep mode	0.0005	5000

$$I_{av} = \sum_{i=1}^{N} I_i D_i + I_{sleep}\left(1 - \sum_{i=1}^{N} D_i\right), \tag{9.1}$$

with $D_i = \Delta t_i / T_{cycle}$, where T_{cycle} is the total duration of each consumption cycle, N is the number of consumption sub-phases per cycle, I_{sleep} is the total current consumption of the system in sleep mode, and I_i and Δt_i are the current consumption and duration of each consumption sub-phase, respectively. The battery lifetime is given by $T_{bat} = \sigma_{IR} C_{bat} / I_{av}$, where C_{bat} is the battery capacity given in ampere-hours, I_{av} is the average current given in amperes (9.1), and σ_{IR} accounts for battery capacity degradation [24]. We consider a simple scenario where the radio transceiver wakes up every T_{cycle}, performs a reception of duration T_{RX} and consumption I_{RX} followed by a transmission of duration T_{TX} and consumption I_{TX}, and then returns to sleep mode for a time T_{sleep} with consumption I_{sleep}.

For $T_{RX} = T_{RX} = T_{ON}$ and $I_{RX} = I_{RX} = I_{ON}$ the average current consumption is $I_{av} = I_{ON}D_{ON} + I_{sleep}(1 - D_{ON})$. The duty cycle $D_{ON} = T_{ON}/T_{cycle}$ represents the fraction of time the main radio is on within a cycle period T_{cycle}, with 1 corresponding to always on and 0 corresponding to always off. See Table 9.1 at the end of this chapter for typical values used in the duty cycle MAC protocol of [22].

9.5.2   Wake-Up Protocol Battery Lifespan

Equation (9.2) gives an estimate of the power consumption and battery lifetime of the wake-up radio scheme proposed in this work as a function of the number of effective wake-ups N (wake-up address successfully decoded) and false alarms M (MCU unintendedly woken up, wake-up address not successfully decoded). Figure 9.5 illustrates a 24-hour consumption cycle with three effective wake-ups of the main radio transceiver and two unintended wake-ups of the wake-up MCU. In the latter case, the main radio remains asleep. This situation is accounted for by the parameter M in (9.2),

$$T_{bat} = \sigma_{IR} C_{bat} \left(\begin{array}{l} I_{dec}\dfrac{\Delta t_{dec}}{T_{cycle}}(N+M) + I_{linkRx}\dfrac{\Delta t_{linkRx}}{T_{cycle}}N + I_{linkTx}\dfrac{\Delta t_{linkTx}}{T_{cycle}}N \\[3mm] + I_{sleep}\left(1 - \dfrac{\Delta t_{dec}}{T_{cycle}}(N+M) - \dfrac{\Delta t_{linkRx}}{T_{cycle}}N - \dfrac{\Delta t_{linkTx}}{T_{cycle}}N\right) \end{array} \right)^{-1}, \tag{9.2}$$

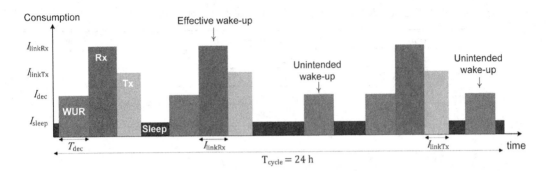

Figure 9.5 Example of 24-hour cycle with five wake-up requests and two false alarms.

where C_{bat} is the battery capacity, I_{sleep} is the sleep mode current consumption (phase 4 in Figure 9.3), I_{dec} and Δt_{dec} are respectively the current consumption and duration of the MCU address decoding (phase 1 in Figure 9.3), I_{linkRX} and Δt_{linkRX} are the current consumption and duration of the main radio reception, respectively (phase 2 in Figure 9.3), and I_{linkTX} and Δt_{linkTX} are the current consumption and duration of the main radio transmission, respectively (phase 3 in Figure 9.3). The parameters N and M account for the total number of effective wake-ups and false alarms per day, respectively. See Table 9.2 at the end of this chapter for typical values used in this analysis.

9.5.3   MAC versus Wake-Up Radio

Figure 9.6(a) compares the power consumption and latency of the MAC and wake-up radio protocols as a function of the duty cycle, as given by (9.1) and (9.2). In this example, it is assumed that a single access to the wireless sensor is expected per day and only the sleep, transmission, reception, and wake-up decoding times are considered in the calculation of the latency (this analysis does not consider additional processing or networking times).

From Figure 9.6(a) we see that (i) the extra power consumption introduced by the wake-up radio MCU and front end leads to higher consumption at very low duty cycles for the wake-up radio protocol, with the turning point at 0.06 % duty cycle, but at very low duty cycles information latency in the MAC protocol can be excessively large; (ii) for large duty cycles, the MAC protocol presents considerably larger consumptions. Moreover, it produces unnecessarily large amounts of traffic for the sporadic sensor access scenario considered here.

Figure 9.6(a) presents the optimal operation region for the wake-up protocol for this example, corresponding to simultaneous low power consumption and latency. The optimal region can be extended by reducing the wake-up power consumption and time (reduced latency). Note that lower power consumption parts like the ATtiny85 MCU and MAX9119 comparator could be used to further improve the wake-up radio gain.

Figure 9.6(b) presents the battery lifetime as a function of the total number of wake-ups and false alarms (i.e., only MCU awake) per day. For a small number of accesses

Table 9.2 Parameters of the proposed wake-up radio protocol.

Parameter	Value	Description
C_{bat}	620 mAh	Battery capacity
σ_{IR}	0.65	Accounts for the battery capacity degradation due to, for example, internal battery resistance and self-discharge
I_{ref}	0.55 μA	Consumption of the comparator voltage reference
$I_{comp_{quiesc}}$	2 μA	Comparator quiescent power consumption
$I_{comp_{active}}$	4 μA	Comparator active power consumption
$I_{WUR_{quiesc}}$	2.55 μA	Wake-up front end quiescent consumption, $I_{ref} + I_{comp_{quiesc}}$
$I_{WUR_{active}}$	4.45 μA	Wake-up front end active consumption, $I_{ref} + I_{comp_{active}}$
$I_{TxRxSleep}$	0.5 μA	Sleep mode consumption of the main transceiver
$I_{uCSleep}$	0.3 μA	ATmega16 power-down consumption with watchdog timer disabled [20]
I_{sleep}	3.85 μA	Sleep mode power consumption
$I_{uCActive}$	3 mA	ATmega16 active mode consumption $V_{DD} = 3$ V and 4 MHz clock [20]
$I_{RxActive}$	31 mA	Main transceiver Rx consumption
$I_{TxActive}$	29 mA	Main transceiver Tx consumption
T_{cycle}	24 hours	Cycle period
$WUR_{BitRate}$	9.6 kbps	Data rate of wake-up pattern
$WUR_{PatternLength}$	6 bytes	Wake-up pattern length
T_{dec}	5.3 ms	Total duration of the wake-up pattern processing including RS232 data reception, CRC check, and address decoding
T_{linkRx}	1.6 ms	Duration of the main link packet reception
T_{linkTx}	1.12 ms	Duration of the main link packet transmission

The consumptions of the various phases defined in Figure 9.3 are as follows:

Phase 1: $I_{dec} = I_{TxRxSleep} + I_{WUR_{active}} + I_{uCActive}$

Phase 2: $I_{linkRX} = I_{RxActive} + I_{WUR_{quiesc}} + I_{uCActive}$

Phase 3: $I_{linkTX} = I_{TxActive} + I_{WUR_{quiesc}} + I_{uCActive}$

Phase 4: $I_{sleep} = I_{TxRxSleep} + I_{WUR_{quiesc}} + I_{uCSleep}$

where $I_{TxRxSleep}$, $I_{RxActive}$, and $I_{TxActive}$ are the current consumptions of the main transceiver in sleep, receiving, and transmitting modes, respectively, $I_{WUR_{active}}$ and $I_{WUR_{quiesc}}$ are the active and quiescent consumptions of the wake-up radio front end, respectively, and $I_{uCSleep}$ and $I_{uCActive}$ are the consumptions of the MCU in sleep and active modes, respectively.

per day, the wake-up radio allows for a maximum lifetime of up 14 years[1] under the conditions specified in Table 9.2. (The lifetime under real conditions may be shorter due to non-ideal conditions not accounted for here.) The number of false alarms, M, seems to have little impact in our example.

Figure 9.6(c) compares the battery lifetime of the wake-up radio protocol and the duty cycle MAC protocol for 5-second, 20-second, and 60-second sleep times and considering five main radio messages per cycle. The crossover points between the two

---

[1] The system may present a shorter battery lifetime due to other non-idealities not captured by the model.

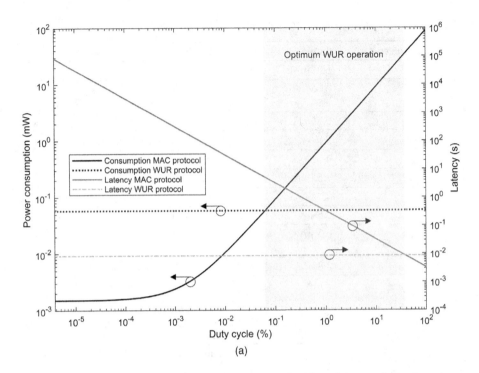

Figure 9.6 (a) Power consumption and latency as a function of duty cycle for the wake-up radio and MAC protocols. The shaded area corresponds to the optimum operation regime for this WUR scheme. (b) Estimated battery lifespan as a function of the number of actual wakeups (N) and false alarms (M). (c) Proposed wake-up protocol versus duty cycle MAC protocol used in [22] for a CR2450 battery (620 mAh capacity). For this example, we assume no false alarms ($M = 0$). The peak battery lifetime estimates obtained here represent best-case scenarios; certain performance degradation mechanisms not accounted for here may reduce these figures.

protocols occur at a very large number of wake-ups or very large MAC sleep time (corresponding to low duty cycle and high latency, respectively). This clearly shows the benefits of the wake-up radio protocol for low-traffic applications.

9.6   Conclusions

To validate the described wake-up radio concept, we designed and prototyped a wake-up radio receiver operating at 860 MHz and integrated it into a wireless sensor node as in Figure 9.2. Our wake-up radio was powered by a 650 mAh lithium battery, exhibited a power consumption of about 10 µW, sensitivity of −35 dBm, and communication range of about 4 m for a transmitted EIRP of 20 dBm. The prototype featured adjustable wake-up sensitivity via a variable voltage reference, an MCU JTAG programming interface, and an 860 MHz dipole antenna (see Figure 9.2). The proposed addressing scheme proved effective in suppressing in-band interference that can trigger false alarms on the main radio, and it can be combined with a preamble

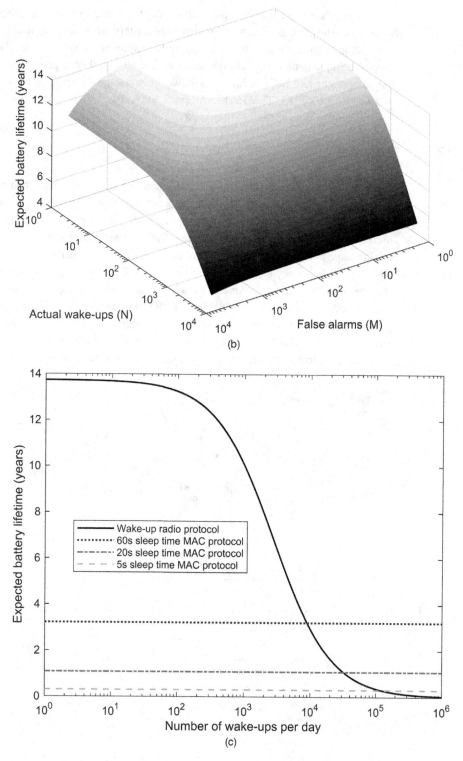

(b)

(c)

Figure 9.6 (cont.)

validation scheme like the one discussed in [25] to further suppress unintended MCU wake-ups. The total power consumption (and battery lifetime) in a low data traffic application is typically dominated by the consumption of the wake-up receiver. In the proposed implementation, this consumption could be further improved by replacing the comparator and MCU with lower-power parts (some examples include AS1976, MAX9119, LPV7215, TLV3691, ATtiny85, MSP430).

9.7   Exercise

Here, we invite you to explore an enhanced interference suppression scheme based on subcarrier modulation (Figure 9.7) that can achieve improved performance compared to conventional implementations. In this approach, the baseband wake-up signal is first used to modulate an intermediate frequency (IF) sub-carrier at frequency $\omega_{subcarrier}$ (first mixing stage in Figure 9.7(a)) and the resulting signal is then used to modulate the main RF carrier at frequency ω_{RF} (second mixing stage). The incoming signal at the receiver (Figure 9.7(b)) is down-converted by a first stage RF envelope detector and then digitized through a comparator, yielding an IF signal centered at $\omega_{subcarrier}$. This signal goes through a resonant filter tuned to $\omega_{subcarrier}$ whose output then goes into a second envelope detector that further down-converts the received IF wake-up signal to baseband. In its simplest implementation, this scheme can be used to awaken the MCU of

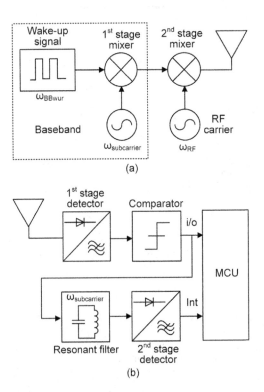

(a)

(b)

Figure 9.7 Improved wake-up scheme with sub-carrier modulation. (a) Transmitter. (b) Receiver.

the wake-up radio by modulating the transmitted sub-carrier with a constant amplitude that creates a positive transition at the output of the second envelope detector. Using a narrowband bandpass filter tuned to the RF carrier frequency at the first stage detector and tuning the resonant filter to the sub-carrier frequency can dramatically improve immunity to interference. In more advanced implementations, this method can be combined with the addressing scheme described previously and can also enable the implementation of multiple wake-up channels by using N sub-carriers (centered at $\omega_{subcarrier1}$, $\omega_{subcarrier2}$, ..., $\omega_{subcarrierN}$) and N resonators tuned to these frequencies.

The goal of this exercise is to implement and validate this wake-up radio scheme. A key element of the proposed system is the resonant filter between the comparator and the second stage detector (see Figure 9.7). Start by simulating the circuit using a square-wave voltage source to mimic the output of the first stage detector (IF wake-up signal), a high-gain operational amplifier as the comparator, and a high quality factor LC resonant circuit. For the second stage detector, use a voltage doubler circuit. Start by plotting the output of the voltage doubler circuit over frequency (Venv_detect_2 in Figure 9.8). To evaluate the frequency selectivity of the circuit, keep the IF resonance frequency fixed and vary the frequency of the sub-carrier applied to the comparator.

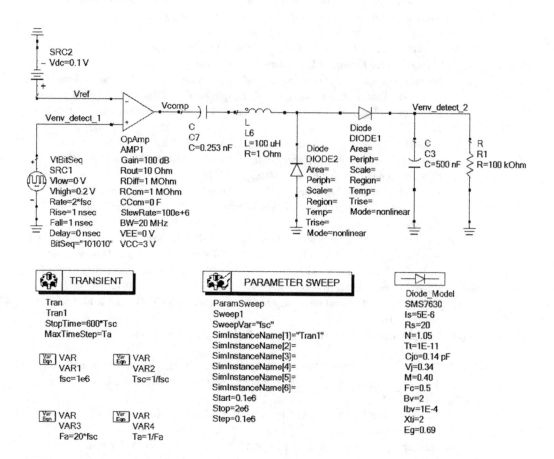

Figure 9.8 Schematic of basic wake-up receiver circuit.

For example, using a 0.253 nF capacitor and 100 µH inductor, a prominent resonance should be seen around 1 MHz. To evaluate the sensitivity of the circuit, plot the time-domain signal of the voltage doubler for a 1 MHz sub-carrier. Lab experiments include evaluating the immunity to interference of this technique and comparing it to the conventional wake-up radio approach.

References

[1] A. Förster, *Emerging Communications for Wireless Sensor Networks*. InTech, Rijeka, Croatia, 2010.
[2] H. Rashidzadeh et al., "Energy Harvesting for IoT Sensors Utilizing MEMS Technology," in *IEEE Canadian Conference on Electrical and Computer Engineering*, 2016, pp. 1–4.
[3] Austria Microsystems, "AS3931 IC." [Online], available at: https://ams.com
[4] Microchip, "ATA5282 IC, MCP2030 IC." [Online], available at: www.microchip.com
[5] EM Microelectronic, "EM4083 IC." [Online], available at: www.emmicroelectronic.com
[6] M. Tomasz, "A Unique, Ultra-low Power Analog IC Enables RF Wakeup Applications," Touchstone Semiconductor, Inc.
[7] D. Benoît et al., "Asynchronous Wake-Up Scheme for Wireless Light Curtains," in *Wireless Congress: Systems and Applications*, 2008.
[8] Impinj. [Online], available at: www.impinj.com
[9] J. Ansari, D. Pankin, and P. Mahonen, "Radio-Triggered Wake-Ups with Addressing Capabilities for Extremely Low Power Sensor Network Applications," in *IEEE 19th International Symposium on Personal, Indoor and Mobile Radio Communications*, 2008.
[10] B. Van der Doorn, W. Kavelaars, and K. Langendoen, "A Prototype Low-Cost Wakeup Radio for the 868 MHz Band," *International Journal of Sensor Networks*, 5(1):22–32, 2009.
[11] P. Le-Huy and S. Roy, "Low-Power 2.4 GHz Wake-Up Radio for Wireless Sensor Networks," in *IEEE International Conference on Wireless and Mobile Computing, Networking and Communications*, 2008, pp. 13–18.
[12] J. M. Lebreton et al., "An Energy-Efficient Duty-Cycled Wake-Up Radio Protocol for Avoiding Overhearing in Wireless Sensor Networks," *Wireless Sensor Network*, 8 (8):176–190, 2016.
[13] I. Haratcherev, M. Fiorito, and C. Balageas, "Low-Power Sleep Mode and Out-Of-Band Wake-Up for Indoor Access Points," in *IEEE Globecom Workshops*, 2009, pp. 1–6.
[14] H. Hong, Y. Kim, and R. Kim, "A Low-Power WLAN Communication Scheme for IoT WLAN Devices Using Wake-Up Receivers," *Appl. Sci.* 8(1):72, 2018.
[15] M. Park et al., "Proposal for Wake-Up Receiver (WUR) Study Group." IEEE 802.11-16/0722r1, 2016.
[16] D. K. McCormick, "Preview," in *IEEE Technology Report on Wake-Up Radio*, 2017, pp. 1–11.
[17] U. A. Perez, "Low Power WiFi: A Study on Power Consumption for Internet of Things," Thesis, Facultat d'Informàtica de Barcelona (FIB) Universitat Politècnica de Catalunya (UPC), 2015.
[18] Maxim Integrated, "MAX9021 Comparator." [Online], available at: www.maximintegrated.com
[19] Microchip, "ATmega16 MCU." [Online], available at: www.microchip.com

[20] Broadcom. [Online], available at: www.broadcom.com

[21] L. Bras, M. Oliveira, N. Borges Carvalho, and P. Pinho, "Low Power Location Protocol based on ZigBee Wireless Sensor Networks," in *International Conference on Indoor Positioning and Indoor Navigation*, 2010, pp. 1–7.

[22] Texas Instruments, "CC2430 System-on-Chip Solution for 2.4 GHz IEEE 802.15.4 / ZigBee." [Online], available at: www.ti.com

[23] K. Aamodt "CC2431 Location Engine," Texas Instruments Application Note AN042, 2006.

[24] Energizer, "Lithium Coin Handbook and Application Manual." [Online], available at: www.energizer.com

[25] M. Magno, V. Jelicic, B. Srbinovski, V. Bilas, E. Popovici, and L. Benini, "Design, Implementation, and Performance Evaluation of a Flexible Low-Latency Nanowatt Wake-Up Radio Receiver," *IEEE Transactions on Industrial Informatics*, 12(2):633–644, 2016.

10 Unconventional Wireless Power Transmission

10.1 Introduction

Wireless power transfer has traditionally been implemented using CW signals, and energy-harvesting circuits have typically been optimized at the circuit level. But recent work [1–6] has investigated an alternative approach for improving the efficiency of wireless power transfer systems through the optimization of wireless power waveforms. This work has shown that signals with high PAPR such as multisine signals can potentially drive energy-harvesting circuits more efficiently than CW signals.

Researchers have explored various alternative waveforms for wireless power transmission, including chaotic signals, ultra-wideband signals, multisine signals, harmonic signals, modulated signals, and white noise [1–16]. The use of multisines gained the most traction and has become the most prominent approach being applied to RFID interrogators [2, 6], medical devices [17, 18], wake-up radios [19], localization systems [20], and novel encoding and modulation schemes [21–23].

In this chapter, we report on various experiments exploiting high-PAPR multisine waveforms to improve wireless power transfer efficiency. The first set of experiments evaluates high-PAPR multisine signal transmission in conducted media, where the energy-harvesting circuit is directly wired to the signal source. The second set of experiments evaluates (over-the-air) wireless power transfer using high-PAPR multisine signals and explores spatial power-combining techniques to efficiently generate and radiate these signals.

10.2 Multicarrier Waveform Design

A multisine signal consists of a sum of N harmonically related subcarriers (10.1), which may interfere constructively or destructively depending on their relative phases:

$$S_{\text{ms}}(t) = \text{Re}\left\{ \sum_{n=1}^{N} V_n e^{j[(\omega_{\min}+(n-1)\Delta\omega)t+\varphi_n]} \right\}, \tag{10.1}$$

Table 10.1 Multisine parameters.

Parameter	Definition	Description
B_ω (Hz)	$(N-1)\Delta\omega$	Total signal bandwidth
V_{peak} (V)	NV_n	Peak amplitude
P_{peak} (Watt)	$N^2 V_n^2 / R_L$	Peak power
P_{AV} (Watt)	$NV_n^2 / 2R_L$	Average power
PAPR_{MAX} (dB)	$10 \log_{10}(2N)$	Peak to average power ratio

Where V_n, $\omega_n = \omega_{\min} + (n-1)\Delta\omega$, and φ_n are the amplitude, frequency, and phase of the nth subcarrier, respectively, N is the number of subcarriers, and $\Delta\omega$ is the subcarrier frequency spacing. Table 10.1 presents the main parameters of a multisine signal. Maximum PAPR is achieved when subcarriers with a constant phase progression interfere constructively (this will be covered in more detail later).

Figure 10.1(a) shows the time-domain waveforms of a CW signal and a four-tone multisine signal with a random phase distribution and the same average power as the CW. Figure 10.1(b) corresponds to phase-aligned subcarriers, which yields maximum PAPR. The corresponding frequency spectra are illustrated in Figure 10.1(c).

10.2.1 Waveform Optimization

In multicarrier wireless power transfer, we are particularly interested in optimizing the signal's bandwidth, subcarrier spacing, and PAPR. In wired experiments, the PAPR produced by the source can be effectively delivered to the energy-harvesting circuit wired to the source. However, in real applications the transmitted PAPR may be altered due to multipath fading and frequency selectivity in the radio channel. To assure conservation of the transmitted PAPR, the bandwidth of the multisine signal should be limited to the channel coherence bandwidth [12].

On the other hand, the subcarrier spacing determines the multisine time-domain peak repetition rate, and excessively low peak repetition rates can degrade the power transfer efficiency. Therefore, optimal subcarrier spacing and bandwidth that provide sufficiently fast peak repetition rate without exceeding the channel coherence bandwidth should be sought [12].

10.2.2 Optimal Phase Distribution

To determine the optimal phase distribution that yields the maximum DC output in an energy-harvesting circuit, consider a rectifying diode whose non-linear conductance can be approximated by a simple polynomial expansion,

$$i_d = k_0 + k_1 v_d + k_2 v_d^2 + k_3 v_d^3 + \cdots + k_n v_d^n, \tag{10.2}$$

where v_d is the voltage across the diode and $k_0, k_1, \ldots, k_n$ are the coefficients of the expansion. Consider a four-tone multisine signal with subcarrier amplitudes

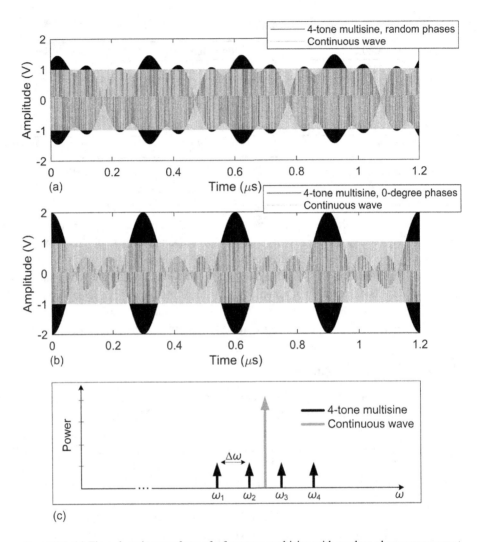

Figure 10.1 (a) Time-domain waveform of a four-tone multisine with random phase arrangement. (b) Four-tone multisine with 0° phase arrangement (black signal) overlaid with CW signal with same average power (gray signal). (c) Illustration of the frequency spectra of the CW and multisine.

$V_1 = V_2 = V_3 = V_4 = V_{ms}$, phases $\varphi_1, \varphi_2, \varphi_3$, and φ_4, and evenly spaced frequencies $\omega_1, \omega_2 = \omega_1 + \Delta\omega, \omega_3 = \omega_1 + 2\Delta\omega$, and $\omega_4 = \omega_1 + 3\Delta\omega$:

$$v_d(t) = V_{ms}\cos(\omega_1 t + \varphi_1) + V_{ms}\cos(\omega_2 t + \varphi_2) + V_{ms}\cos(\omega_3 t + \varphi_3) + V_{ms}\cos(\omega_4 t + \varphi_4).$$

$$(10.3)$$

Exciting the system in (10.2) truncated to order four with the input signal (10.3) yields the following output DC component:

$$I_{DC}(\varphi_1, \varphi_2, \varphi_3, \varphi_4) = \frac{\Delta\omega}{2\pi N} \int\limits_{-\pi N/\Delta\omega}^{\pi N/\Delta\omega} i_d(t) \; dt$$

$$\approx k_0 + \left[\begin{array}{l} 4V_{ms}^2 k_2 + 21V_{ms}^4 k_4 + 3V_{ms}^4 k_4 \cos(2\varphi_3 - \varphi_2 - \varphi_4) \\ +3V_{ms}^4 k_4 \cos(-2\varphi_2 + \varphi_1 + \varphi_3) + 6V_{ms}^4 k_4 \cos(\varphi_1 - \varphi_2 - \varphi_3 + \varphi_4) \end{array} \right] \bigg/ 2,$$

$$\tag{10.4}$$

where the integral operator represents an ideal low-pass filter used to eliminate high-frequency components produced by the rectifier non-linearity. The phase-dependent term in (10.4) can be maximized by using a constant phase progression of the form $\varphi_{i+1} - \varphi_i = \Delta\varphi$, with $\Delta\varphi$ being a constant phase shift between consecutive subcarriers. By phase-normalizing the second, third, and fourth subcarriers to the first, the optimal phase progression becomes $\varphi_2 = \varphi_1 + \Delta\varphi$, $\varphi_3 = \varphi_1 + 2\Delta\varphi$, and $\varphi_4 = \varphi_1 + 3\Delta\varphi$, which effectively nulls the arguments of the cosines in (10.4) and maximizes the DC component,

$$\begin{cases} 2(\varphi_1 + 2\Delta\varphi) - (\varphi_1 + \Delta\varphi) - (\varphi_1 + 3\Delta\varphi) = 0, \\ -2(\varphi_1 + \Delta\varphi) + \varphi_1 - (\varphi_1 + 2\Delta\varphi) = 0, \\ \varphi_1 - (\varphi_1 + \Delta\varphi) - (\varphi_1 + 2\Delta\varphi) + (\varphi_1 + 3\Delta\varphi) = 0. \end{cases} \tag{10.5}$$

Note that this is the same condition that yields maximum PAPR in the multisine waveform (10.1) and leads to phase mode-locking in active antenna arrays (more on this in Section 10.5.3). The most trivial optimal phase progression corresponds to having all subcarriers with $0°$ relative phase. Besides the phases, the amplitudes of the subcarriers can also be optimized [3].

10.2.3 Multisine Gain

To evaluate the multisine efficiency gain in the experiments presented in this chapter, we introduce the following figure of merit:

$$G_\eta(dB) = 10\log_{10}\left(\frac{\eta_{ms}}{\eta_{cw}}\right) = 10\log_{10}\left(\frac{P_{dc_{ms}}/P_{rf_{ms}}}{P_{dc_{cw}}/P_{rf_{cw}}}\right), \tag{10.6}$$

where $P_{rf_{cw}}$ and $P_{rf_{ms}}$ are the CW and multisine average input power levels, respectively, $P_{dc_{cw}}$ and $P_{dc_{ms}}$ are the DC power levels collected under CW and multisine excitations, respectively, and η_{cw} and η_{ms} are the efficiencies obtained under CW and multisine excitations, respectively.

Specifically, if the input power and output DC load are the same for the CW and multisine excitation ($R_{L_{cw}} = R_{L_{ms}}$ and $P_{rf_{cw}} = P_{rf_{ms}}$), then G_η can be given in terms of the collected DC voltages, $V_{dc_{cw}}$ and $V_{dc_{ms}}$,

$$G_\eta(dB) = 10\log_{10}\left(\frac{P_{dc_{ms}}}{P_{dc_{cw}}}\right) = 10\log_{10}\left(\frac{V_{dc_{ms}}^2}{V_{dc_{cw}}^2}\right). \tag{10.7}$$

A more general definition of RF-to-DC efficiency is given by

$$
\eta = \frac{P_{\text{out}}}{P_{\text{in}}} = \frac{\displaystyle\int_{-T/2}^{T/2} v_{\text{out}}(t) i_{\text{out}}(t) \; dt}{\displaystyle\int_{-T/2}^{T/2} v_{\text{in}}(t) i_{\text{in}}(t) \; dt},
\tag{10.8}
$$

where $v_{\text{in}}(t)$ and $i_{\text{in}}(t)$ are the input time-domain voltage and current, respectively, $v_{\text{out}}(t)$ and $i_{\text{out}}(t)$ are the time-domain voltage and current at the output DC load, respectively, and T is the period of the CW carrier or multisine envelope.

10.3   Wired Experiments

In this section, we evaluate the multisine efficiency gain of two energy-harvesting circuits directly wired to an RF signal generator. We tested a single HSMS2850 diode rectifier, which was optimized to operate at 2.3 GHz [4, 24], and a five-stage charge pump based on HSMS2852 diode pairs, which was optimized to work at 866.6 MHz [25, 26]. The charge pump is part of the RF front end of a passive wireless sensor developed in [25]. This front end, whose schematic is depicted in Figure 10.2, also comprises backscatter modulation circuitry (bottom left of Figure 10.2), storage capacitor, over-voltage protection, and power management circuitry. For further details on the design and implementation of this passive wireless sensor node, refer to [25].

The two energy-harvesting circuits under test were excited with a CW signal and various multisine signals with power levels ranging from -30 dBm to 0 dBm. For maximum PAPR, the relative phase of each multisine subcarrier was set to $0°$. For this experiment, the DC output of the charge pump in Figure 10.2 was isolated from the rest of the passive wireless sensor circuit and was loaded with a 510 kΩ resistor.

Figures 10.3 and 10.4 present the measured results showing multisine efficiency gains of up to 6 dB and 2.7 dB for the 2.4 GHz rectifier and 866.6 MHz charge pump circuits, respectively. The efficiency gain increased with increasing number of tones (and PAPR), except for the 64-tone multisine signal. The efficiency drop in the 64-tone case is believed to be due to saturation of the signal generator at high PAPR and power loss resulting from the limited bandwidth of the rectifier circuit.

10.4   Passive Wireless Sensor Experiments

In this experiment, the charge pump circuit was connected to the input DC supply of the passive wireless sensor and was first driven with a CW signal at 866.6 MHz and then with a four-tone multisine signal with subcarriers at 865.7 MHz, 866.3 MHz,

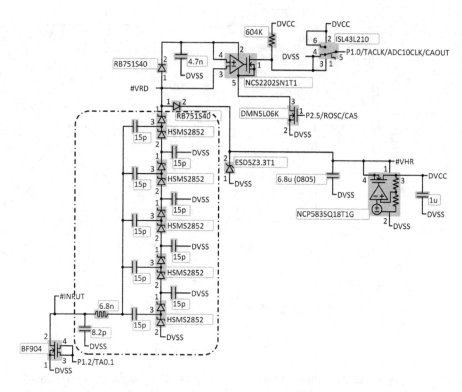

Figure 10.2 Passive wireless sensor front end tested under multisine excitations. The dash-dotted line encloses the energy-harvesting circuit used in these experiments. Figure reprinted from [25].

866.9 MHz, and 867.6 MHz. We varied the average power level from −15 dBm to −1 dBm and then back to −15 dBm.

As seen from Figure 10.5, the four-tone multisine provides a higher DC voltage and lower turn-off threshold compared to the CW signal (the sensor's turn-on and turn-off power thresholds are represented by the vertical solid lines in Figure 10.5). These results also suggest that this passive wireless sensor has hysteretic behavior.

10.4.1 Communication Range Gain

To evaluate the potential communication range gains that can be achieved with multisine signals, the passive wireless sensor of [25] was first illuminated with a CW signal and then with a four-tone multisine signal with the same average power. Both signals were created with a signal generator, amplified using an external power amplifier, then radiated through a circularly polarized antenna.

The radiated signal was harvested by the passive wireless sensor several meters away from the transmit antenna using a half-wavelength dipole antenna. A light-emitting diode (LED) in the passive wireless sensor node was used to detect the maximum operating range under CW and multisine excitations – the sensor's

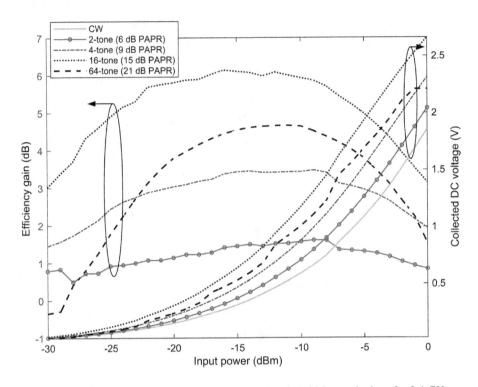

Figure 10.3 DC voltage and efficiency gain under CW and multisine excitations for 2.4 GHz single-diode energy-harvesting circuit.

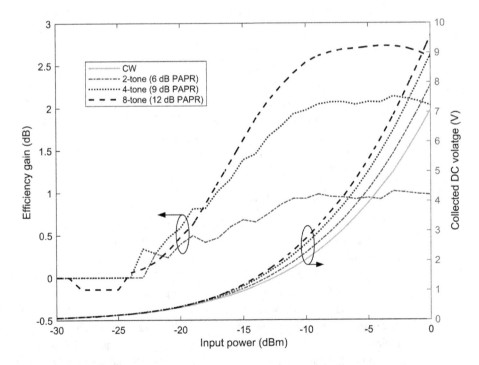

Figure 10.4 DC voltage and efficiency gain under CW and multisine excitations for 866.6 MHz five-stage energy-harvesting circuit.

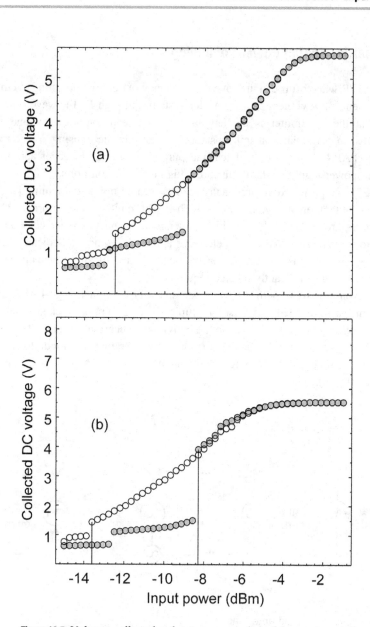

Figure 10.5 Voltages collected at the storage capacitor of the passive wireless sensor as a function of increasing and decreasing input power levels (filled and empty circles, respectively). (a) Wireless sensor illuminated with a CW signal. (b) Wireless sensor illuminated with a four-tone multisine signal. Results from [26].

microcontroller was programmed to flash the LED when the harvested power was enough to activate the sensor. By moving the passive wireless sensor away from and towards the wireless power transmission antenna, the maximum distance was found to be 4.1 m for the CW signal and 5.3 m for the multisine signal, suggesting an improvement of about 30%.

10.5   Multicarrier Spatial Power Combining

High-PAPR waveforms can improve the efficiency of energy-harvesting circuits at the receiver end of a wireless power transfer system, but high-PAPR waveform amplification at the transmitter is challenging and can lead to efficiency and linearity problems. In particular, transmitter saturation can clip the transmit waveform, reducing its PAPR, which would defeat the purpose of the multicarrier approach. The transmit power amplifier should feature a high compression point.

In [27], we proposed an approach based on spatial power combining [28, 29] for efficient generation and radiation of high-PAPR multisine waveforms, where the subcarriers are individually amplified and radiated, and passively combine in free space to form a high-PAPR EM field. This approach requires a dedicated generator, amplifier, and antenna for each subcarrier, but each amplification stage operates in the CW mode with significantly relaxed requirements.

We explored two schemes to generate spatially combined high-PAPR multisine fields. In the first scheme, we used multiple signal generators locked to an external 10 MHz phase reference to generate the individual subcarriers (Figure 10.6(a)). In the second scheme, we used an active antenna array operating in a mode-locked regime that requires no external phase reference to create phase-locked multisine subcarriers (Figure 10.6(b)).

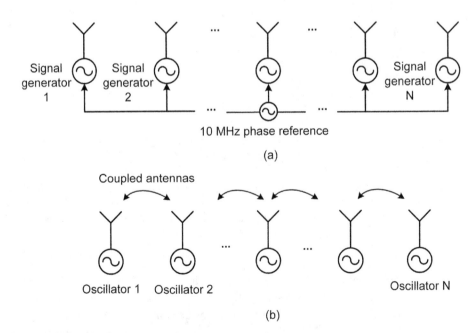

Figure 10.6 (a) Externally synchronized transmitter. (b) Mode-locked transmitter.

10.5.1   Over-the-Air Subcarrier Combining

Spatial power combining has been applied at millimeter-wave frequencies to achieve high output power by combining the outputs of several low-power amplifiers over the air [28, 29]. Assuming far-field observation [30], small separation between adjacent transmit antenna elements, and flat channel frequency response, the combined electrical field of multiple antenna elements at a distance r and angle θ can be approximated by

$$E(\theta, r) = \sum_{n=1}^{N} E_n G_n(\theta) e^{j[\omega_n t + \gamma r + \phi_n]},$$ (10.9)

where E_n, ω_n, ϕ_n, and $G_n(\theta)$ are the field amplitude, frequency, phase, and angle-dependent gain of each transmit antenna element n, respectively, and γ is the propagation constant.

To maximize the PAPR of the spatially combined electrical field, an evenly spaced subcarrier frequency grid ($\omega_{n+1} - \omega_n = \Delta\omega$ with $\Delta\omega$ constant) and a constant subcarrier phase progression ($\phi_{n+1} - \phi_n = \Delta\phi$ with $\Delta\phi$ constant) are required [29]. Note that these are the same conditions that yielded maximum DC current in (10.4).

10.5.2   Spatially Combined Multisine Transmitter

We used three signal generators locked to an external 10 MHz phase reference to generate three subcarriers in the UHF band, which were radiated through three separate antennas to produce a spatially combined three-tone multisine field (Figure 10.7). To harvest the spatially combined multisine, we used a five-stage

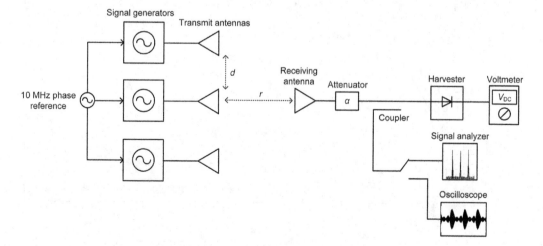

Figure 10.7 Diagram of the measurement setup to evaluate spatial power combining. Left: Spatially combined multisine transmitter with external phase reference. Right: Energy-harvesting receiver circuit. $d = 9$ cm, $r = 70$ cm.

charge pump circuit like the one described in Section 10.3 attached to a dipole antenna and placed 70 cm away from the transmit antennas.

One of the signal generators was used as the phase reference to which the other two were locked via its 10 MHz output clock signal. To guarantee stable phase-locking, we used short identical cables for distributing the 10 MHz phase reference signal. In addition, we monitored the signal at the input of the energy-harvesting circuit to ensure the sources stayed locked during the experiment. This was achieved by measuring the PAPR of the received signal using a directional coupler and an oscilloscope, as in Figure 10.7.

The frequencies of the three subcarriers were set to $f_1 = 876$ MHz, $f_2 = 877$ MHz, and $f_3 = 878$ MHz. For comparison, we also illuminated the energy-harvesting circuit with a CW signal at $f_{CW} = 877$ MHz. The CW and multisine average power levels at the input of the energy-harvesting circuit were set to the same value by adjusting the attenuator at the input of the circuit. The average power level was measured using a directional coupler and a spectrum analyzer (see Figure 10.7).

To evaluate the gain provided by the spatially combined multisine, we used the efficiency gain figure of merit defined in (10.5). Figure 10.8 shows the results obtained in the experiment of Figure 10.7 for locked and unlocked subcarrier phases. To unlock

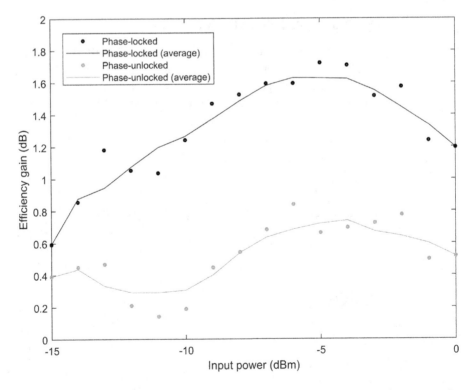

Figure 10.8 Multicarrier RF-to-DC conversion efficiency as a function of average input power for synchronized and unsynchronized subcarriers. The average input power was measured at the input of the energy-harvesting circuit using a coupler and a spectrum analyzer.

the subcarriers, we disconnected the 10 MHz phase reference signal and let the signal generators run freely. As expected, unlocking the subcarriers significantly reduced the gain. For the phase-locked case, which provided the highest PAPR, we obtained a maximum gain of 1.72 dB, corresponding to a DC power increase of 49%.

10.5.3   Mode-Locked Active Antenna Array

Here, we use an active antenna array to efficiently generate spatially combined high-PAPR multisine waveforms [27]. In this scheme, the constant phase progression required for maximum PAPR is achieved via a mode-locking phenomenon occurring in oscillators that are electromagnetically coupled via antennas [31–33]. This phenomenon allows the oscillators to self-lock to each other, and therefore no external phase reference is required.

By controlling the free-running frequencies of each oscillator, one can select the phase distribution of the generated multisine subcarriers. The subcarriers' phase distribution range and the type of signals that can be generated are determined by the complex coupling between the elements of the antenna array.

A 4×1 array of mode-locked antenna oscillators was designed to operate in the C-band [27]. The core element of this is an active antenna oscillator consisting of a patch antenna aperture-coupled to a voltage-controlled oscillator element (see Figure 10.9). The antenna patches were fabricated on 0.5 mm Arlon A25N substrate with dielectric constant 3.38 and loss tangent 0.0025. The active circuitry and coupling slots layer were fabricated in Rogers 4003C (0.5 mm thickness) with dielectric constant 3.38 and loss tangent 0.0027. The two substrates were separated by a 3 mm Rohacell foam layer and the three layers bonded using a 3M spray adhesive. The selected VCO was the commercial Z-COMM 6200L-LF with an output power of approximately 3 dBm.

10.5.3.1   Mode-Locked Multisine Signal Synthesis

To synthesize a mode-locked multisine signal using the 4×1 array, the subcarrier frequency spacing of the four oscillator elements must be large enough that the free-running frequencies of the subcarriers do not synchronize to a common frequency. Initially, only two oscillator elements were turned on, and their frequencies were set to 6.18 GHz and 6.23 GHz. A spacing of 50 MHz was selected to prevent the two oscillators from synchronizing to a common frequency. Because of the two frequency components, mixing products appear at 6.13 GHz and 6.28 GHz.

Afterward, the other two oscillators were turned on one at a time. Setting these oscillators' frequencies close enough to the frequencies of the mixing products at 6.13 GHz and 6.28 GHz causes them to lock to those frequencies, creating a four-tone multisine signal with equally spaced subcarriers. Due to the mode-locking phenomenon, the generated subcarriers have a constant phase progression between them (refer to [27] for more details).

Once the mode-locked state is reached, the free-running frequencies of the oscillator elements of the array can be varied within a certain frequency range to establish different subcarrier phase distributions. Additionally, selecting free-running

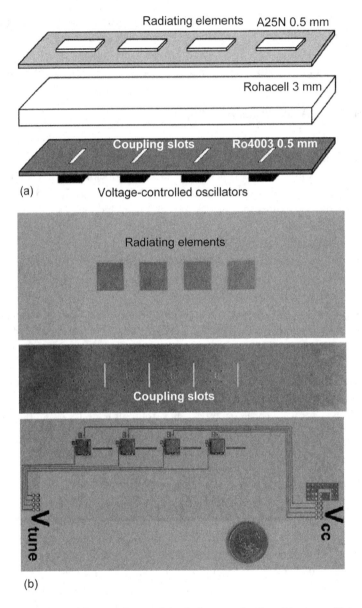

Figure 10.9 (a) Layer scheme of the 4×1 array of active antenna oscillators. (b) Implemented 4×1 antenna array. Reprinted from [27].

frequencies that are closer or farther away from each other allows the synthesis of multisine signals with smaller or larger subcarrier spacings.

The minimum subcarrier spacing is limited by the synchronization bandwidth of the oscillators. This synchronization bandwidth depends on the coupling between the antenna elements. In this experiment, the synchronization bandwidth of the oscillators was measured to be approximately 90 MHz, which means that a subcarrier spacing less

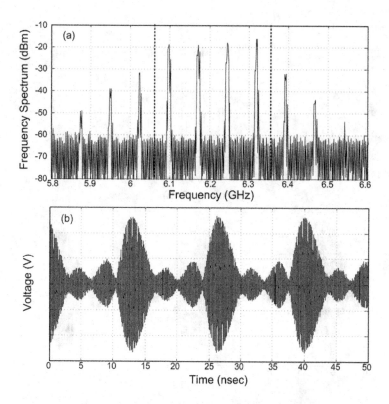

Figure 10.10 Measured mode-locked multisine signal with subcarrier frequency spacing of 75 MHz. (a) Frequency spectrum. (b) Time-domain waveform. Reprinted from [27].

than 45 MHz would cause the oscillators to lock to a common frequency and multisine mode-locking would not be achieved. If a smaller subcarrier frequency spacing is desired, the active antenna oscillator array should be designed to have smaller synchronization bandwidth, which can be done by reducing the coupling between the antenna elements. However, the smaller the synchronization bandwidth, the more sensitive the system is to interference that may affect the mode-locked operation.

Using the 4×1 array, two mode-locked four-tone multisine signals with subcarrier frequency spacing of 75 MHz and 45 MHz were generated. The free-running frequencies of the four oscillators were selected to achieve the desired subcarrier spacing and the constant phase progression that produces maximum PAPR. A PAPR of approximately 8.9 dB was obtained for the signals in Figures 10.10 and 10.11.

10.5.3.2 CW Signal Synthesis
The same 4×1 array was used to synthesize a CW signal for comparison purposes. Unlike the multisine case, to synthesize a CW signal, all the oscillators must synchronize to a common frequency. This was achieved by setting all the oscillators to the same free-running frequency (6.2 GHz). Figure 10.12 depicts the generated CW signal.

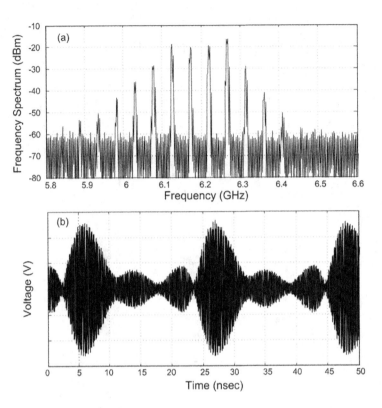

Figure 10.11 Measured mode-locked multisine with subcarrier frequency spacing of 45 MHz. (a) Frequency spectrum. (b) Time-domain waveform. Reprinted from [27].

10.5.3.3   Performance Evaluation

Figure 10.13 illustrates the measurement setup used to evaluate the mode-locked active antenna array and determine the multisine gain on the energy-harvesting circuit. The transmitter consisted of the 4×1 active antenna oscillator array of Figure 10.9, and the receiver comprised a horn antenna and a rectifier circuit based on the MACOM MA2502L Schottky diode. The rectifier circuit, employing an LC input-matching network and an output load ($10 \text{ k}\Omega$), was optimized for maximum RF-to-DC conversion efficiency at 6.2 GHz. The transmitter and the receiver were spaced apart by approximately 30 cm. To simulate the effect of larger distances between the transmitter and receiver, two variable attenuators, α_1 and α_2, with 1 dB and 10 dB steps were used in the receiver between the receiving horn antenna and the rectifier.

During the experiment, the attenuation values α_1 and α_2 were varied and the harvested DC voltage (V_{DC}) was measured using a multimeter. The available power level at the input of the rectifier was measured using a 10 dB directional coupler and an oscilloscope. The power level at the input of the rectifier was averaged across the frequency band that covers the four dominant multisine subcarriers (between the two vertical dashed lines in Figure 10.10).

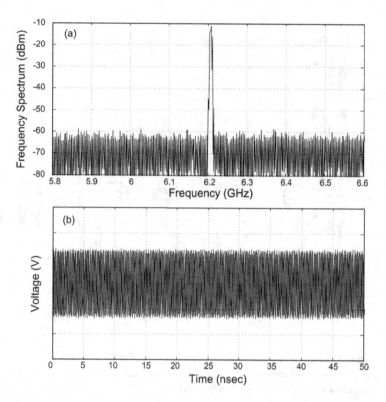

Figure 10.12 Measured CW signal. (a) Frequency spectrum. (b) Time-domain waveform. Reprinted from [27].

A similar procedure was used for the CW signal and the results were compared to the mode-locked four-tone case to determine the multisine efficiency gain. The attenuation values were adjusted to set the power of the CW signal to the same level as the four-tone mode-locked multisine signal. Figure 10.14 depicts the multisine efficiency gain as a function of the average input power level for a mode-locked four-tone multisine with 45 MHz and 75 MHz subcarrier frequency spacings.

10.6   Conclusions

In this chapter, we discussed wireless power transfer using unconventional signals with a focus on multicarriers, provided guidelines for multicarrier waveform optimization, and reported on various experiments including wired circuit-level experiments and over-the-air experiments.

The wired experiments showed gains of up to 6 dB and 2.7 dB for a single-diode rectifier and a five-stage charge pump circuit, respectively. We showed a 30% coverage range improvement in a custom passive wireless sensor node illuminated with a four-tone multisine signal. We also proposed an approach for efficient

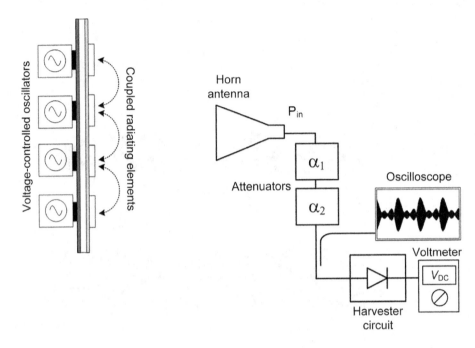

Figure 10.13 Diagram of the measurement setup used to evaluate the multisine gain. The energy-harvesting circuit is illuminated with a spatially combined multisine signal generated by the active antenna array of Figure 10.9. Reprinted from [27].

generation and transmission of high-PAPR multicarrier signals and showed an efficiency gain of up to 15 dB for a 4×1 mode-locked active antenna array.

The main drawbacks of multicarrier wireless power transmission include increased transmitter complexity, increased bandwidths (not currently regulated for wireless power transfer application), and challenging high-PAPR amplification. The latter can be overcome with the space power-combining techniques proposed here.

10.7   Exercises

1. Rewrite the equation for the RF power collected by a transponder [(3.36) in Chapter 3, reproduced below for convenience] to account for the multisine efficiency gain (G_η).

$$P_{tag}[\text{dBm}] = P_{tx} + G_{tx} + G_{tag} + 20 \log_{10}\left(\frac{\lambda}{4\pi d}\right) + 10 \log_{10}(1 - \beta) + \theta_f - F_f$$

$$(10.10)$$

2. Determine the DC power that would be delivered to a transponder via a multicarrier. Hint: Use the equation obtained in Exercise 1 and consider the RF-to-DC conversion efficiency under CW excitation.

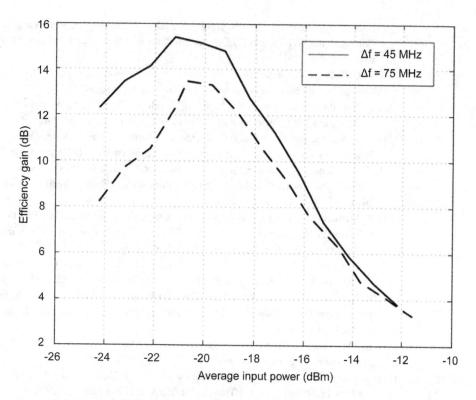

Figure 10.14 Measured RF-to-DC efficiency gain as a function of the average power at the input of the rectifier circuit. Reprinted from [27].

References

[1] H. Matsumoto and K. Takei, "An Experimental Study of Passive UHF RFID System with Longer Communication Range," in *Proceedings of the Asia-Pacific Microwave Conference*, 2007.

[2] M. S. Trotter, J. D. Griffin, and G. D. Durgin, "Power-Optimized Waveforms for Improving the Range and Reliability of RFID Systems," in *IEEE International Conference on RFID*, 2009, pp. 80–87.

[3] M. S. Trotter and G. D. Durgin, "Survey of Range Improvement of Commercial RFID Tags with Power Optimized Waveforms," in *2010 IEEE International Conference on RFID*, 2010, pp. 195–202.

[4] A. S. Soares Boaventura and N. B. Borges Carvalho, "Maximizing DC Power in Energy Harvesting Circuits Using Multisine Excitation," *2011 IEEE MTT-S International Microwave Symposium*, 2011, pp. 1–4.

[5] C.-C. Lo, et al., "Novel Wireless Impulsive Power Transmission to Enhance the Conversion Efficiency for Low Input Power," in *IEEE MTT-S International Microwave Workshop Series on Innovative Wireless Power Transmission*, 2011, pp. 55–58.

[6] A. J. Soares Boaventura and N. Borges Carvalho, "Extending Reading Range of Commercial RFID Readers," *IEEE Transactions on Microwave Theory and Techniques*, 61(1):633–640, 2013.

[7] A. Collado and A. Georgiadis, "Improving Wireless Power Transmission Efficiency Using Chaotic Waveforms," in *International Microwave Symposium*, 2012.

[8] A. Collado and A. Georgiadis, "Optimal Waveforms for Efficient Wireless Power Transmission," *IEEE Microwave and Wireless Components Letters*, 24(5):354–356, 2014.

[9] C. R. Valenta and G. D. Durgin, "Rectenna Performance Under Power-Optimized Waveform Excitation," in *IEEE International Conference on RFID*, 2013, pp. 237–244.

[10] C. R. Valenta, "Microwave-Energy Harvesting at 5.8 GHz for Passive Devices." Ph.D. Thesis, School of Electrical and Computer Engineering, Georgia Institute of Technology, 2014.

[11] A. Litvinenko, J. Eidaks, and A. Aboltins, "Usage of Signals with a High PAPR Level for Efficient Wireless Power Transfer," in *IEEE Sixth Workshop on Advances in Information, Electronic and Electrical Engineering*, 2018, pp. 1–5.

[12] N. Pan, et al., "Multi-Sine Wireless Power Transfer with a Realistic Channel and Rectifier Model," in *2017 IEEE Wireless Power Transfer Conference (WPTC)*, 2017, pp. 1–4.

[13] B. Clerckx and E. Bayguzina, "Low-Complexity Adaptive Multisine Waveform Design for Wireless Power Transfer," *IEEE Antennas and Wireless Propagation Letters*, 16:2207–2210, 2017.

[14] B. Clerckx and E. Bayguzina, "Waveform Design for Wireless Power Transfer," *IEEE Transactions on Signal Processing*, 64(23):6313–6328, 2016.

[15] A. Soares Boaventura et al., "Boosting the Efficiency: Unconventional Waveform Design for Efficient Wireless Power Transfer," *IEEE Microwave Magazine*, 16(3):87–96, 2015.

[16] A. Soares Boaventura et al., "Optimum Behavior: Wireless Power Transmission System Design through Behavioral Models and Efficient Synthesis Techniques," *IEEE Microwave Magazine*, 14(2):26–35, 2013.

[17] H. Zhang et al., "Wireless Power Transfer Antenna Alignment Using Intermodulation for Two-Tone Powered Implantable Medical Devices," *IEEE Transactions on Microwave Theory and Techniques*, 67(5):1708–1716, 2019.

[18] B. Wang and H. Zhang, "Ultra-Wide Dynamic Range Rectifier Topology for Multi-Sine Wireless Powered Endoscopic Capsules," in *2018 IEEE MTT-S International Wireless Symposium*, 2018, pp. 1–4.

[19] F. Hutu and G. Villemaud, "On the Use of the FBMC Modulation to Increase the Performance of a Wake-Up Radio," in *IEEE Radio and Wireless Symposium (RWS)*, 2018, pp. 139–142.

[20] N. Decarli et al., "High-Accuracy Localization of Passive Tags with Multisine Excitations," *IEEE Transactions on Microwave Theory and Techniques*, 66 (12):5894–5908, 2018.

[21] D. I. Kim, J. H. Moon, and J. J. Park, "New SWIPT Using PAPR: How it Works," *IEEE Wireless Communications Letters*, 5(6):672–675, 2016.

[22] M. Rajabi et al., "Modulation Techniques for Simultaneous Wireless Information and Power Transfer with an Integrated Rectifier–Receiver," *IEEE Transactions on Microwave Theory and Techniques*, 66(5):2373–2385, 2018.

[23] S. Claessens, N. Pan, D. Schreurs, and S. Pollin, "Multitone FSK Modulation for SWIPT," *IEEE Transactions on Microwave Theory and Techniques*, 67(5):1665–1674, 2019.

[24] Avago Technologies, "DEMO-HSMS285–0, Demonstration Circuit Board for HSMS-2850, HSMS-2852 and HSMS-2855."

[25] R. D. Fernandes, "Design of a Battery-Free Wireless Sensor Node." M.Sc. thesis, University of Aveiro, 2010. [Online], available at: https://ria.ua.pt/bitstream/10773/5616/1/MSC_RDF.pdf

[26] R. D. Fernandes et al., "Increasing the Range of Wireless Passive Sensor Nodes Using Multisines," in IEEE International Conference on RFID – Technologies and Applications, 2011, pp. 549–553.

[27] A. J. Soares Boaventura, A. Collado, A. Georgiadis, and N. Borges Carvalho, "Spatial Power Combining of Multi-Sine Signals for Wireless Power Transmission Applications," *IEEE Transactions on Microwave Theory and Techniques*, 62(4):1022–1030, 2014.

[28] J. Harvey, E. R. Brown, D. B. Rutledge, and R. A. York, "Spatial Power Combining for High-Power Transmitters," *IEEE Microwave Magazine*, 1(4):48–59, 2000.

[29] R. A. York and R. Compton, "Coupled-Oscillator Arrays for Millimeter-Wave Power-Combining and Mode-Locking," in *IEEE International Microwave Symposium Digest*, Vol. 1, 1992, pp. 429–432.

[30] C. A. Balanis, *Antenna Theory: Analysis Design*, 3rd ed. John Wiley & Sons, Hoboken, NJ, 2005.

[31] R. A. York and R. C. Compton, "Experimental Observation and Simulation of Mode-Locking Phenomena in Coupled-Oscillator Arrays," *Journal of Applied Physics*, 71 (6):2959–2965, 1992.

[32] R. A. York and R. C. Compton, "Mode-Locked Oscillator Arrays," *IEEE Microwave and Guided Wave Letters*, 1(8):215–218, 1991.

[33] R. A. York and R. C. Compton, "Measurement and Modelling of Radiative Coupling in Oscillator Arrays," *IEEE Transactions on Microwave Theory and Techniques*, 41 (3):438–444, 1993.

11 A Battery-less Backscatter Remote Control System

11.1 Introduction

The remote control (RC) is perhaps one of the most widespread convenience features ever invented [1]. But there are major drawbacks relating to their typical source of power, disposable batteries. Besides their limited lifespan and maintenance cost, batteries can produce toxic waste, which translates not only into costs associated with hazardous waste treatment but also environmental costs. In the US alone, the global annual production of alkaline batteries exceeds 10 billion units [2], and in the European economic area, around 10.5 billion portable batteries were placed on the market in 2015 [3]. Moreover, the recent international COVID-19 pandemic has exposed a new problem in large supply chains, including those for non-rechargeable alkaline batteries.

To extend battery lifetime, RC devices typically use efficient wake-up schemes that minimize their overall power consumption. Low-cost RC applications commonly use infrared (IR) wavelengths for data communication. For applications requiring long-range, two-way, non-line-of-sight communication, RF-based RC technology is typically used. To provide a uniform RC wireless standard that enables interoperability between different devices and vendors, the ZigBee alliance has recently introduced the Radio Frequency for Consumer Electronics (RF4CE) standard [4].

Various energy-harvesting approaches, including mechanical energy harvesting [5, 6], have recently been proposed to mitigate the drawbacks of disposable batteries in RCs. In [7], the authors presented a self-powered RC incorporating a piezoelectric push button that can produce enough energy to power up a digital encoder and a low-power RF transmitter when pressed by the user. This kind of approach could enable controls and interfaces to be introduced into interactive environments without requiring wires or batteries.

In this chapter, we describe a new method that eliminates the need for batteries in RC devices. This method uses dedicated wireless power transmission to remotely supply the RC device via an RF power transmitter installed in the controlled device (e.g., a TV receiver). To communicate with the controlled device, the RC reflects the received RF signal with a specific signature to encode messages that can be intercepted and decoded by the controlled device (see the illustration in Figure 11.1). We used off-the-shelf devices to implement an RFID-inspired battery-less RC

Controlled device (e.g., TV set)

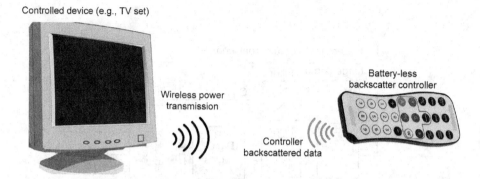

Wireless power
transmission

Battery-less
backscatter controller

Controller
backscattered data

Figure 11.1 Illustration of the proposed battery-less backscatter RC approach. In the prototype presented in this chapter, the controlled device incorporates an RFID interrogator to wirelessly power the RC and decode RF backscattered data. Note: The pictures used here are not from the actual devices and are used for illustration purposes only (for details of the actual implementation, see the following sections).

prototype and we then incorporated this prototype into a commercial TV receiver to showcase a practical eco-friendly application of the wireless power transmission and backscatter concepts to a real home automation scenario.

11.2   The Proposed System

The following describes the architecture of the proposed system and details of its implementation, including the RFID chip control mechanism, custom single-pole, multiple-throw switch, and antenna multiplexing system.

11.2.1   System Architecture

In the proposed system of Figure 11.1, the device being controlled has a built-in RF transceiver to wirelessly power the RC and decode its backscattered signal. Another approach, detailed later in this chapter, consists of using an external RFID interrogator and an interfacing module to link the interrogator to the controlled device via IR.

One way to implement the battery-less RC is to combine button decoding circuitry with a backscatter front end like those found in passive RFID transponders as illustrated in Figure 11.2(a). An alternative approach, illustrated in Figure 11.2 (b), consists of pairing each RC button with an individual RFID chip. In our prototype, we used off-the-shelf RFID chips and a custom single-pole, multiple-throw (SPnT) switch for sharing an antenna between these chips (see Figure 11.2 (b)). Signal routing in the SPnT switch is achieved via user-actuated tactile push-button switches.

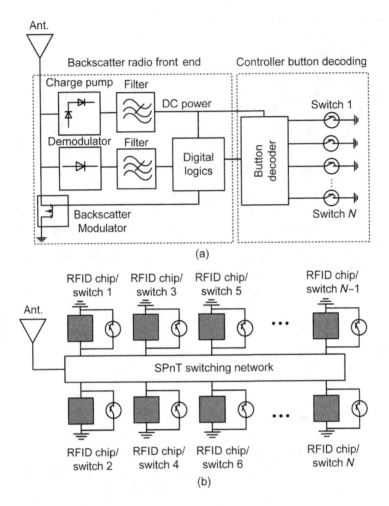

Figure 11.2 (a) RFID-inspired battery-less backscatter RC. (b) Battery-less RC based on multi-RFID scheme incorporating a user-controllable SPnT switch.

## 11.2.2	RFID Chip Control

For the multi-RFID approach in Figure 11.2(b), we require a control mechanism that allows us to activate and deactivate the individual RFID chips at will. One way to implement this is by using a normally closed switch connected in parallel with each RFID chip. In their default position, the switches shunt the terminals of the RFID chips, rendering them inactive and unable to respond to the RFID reader. As the user presses the desired button the switch opens, and the corresponding RFID chip is routed to the RC antenna and can be interrogated by the RFID reader. For this implementation, we used off-the-shelf UHF RFID chips and tactile push-button switches.

To impose a frequency-tuned short-circuit across the terminals of each RFID chip, we connect a capacitor (C_{res}) in series with the switch. This capacitor and the switch's

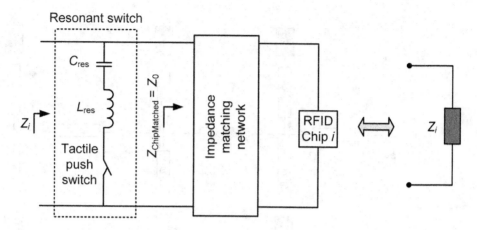

Resonant switch

C_{res}

L_{res}

Tactile push switch

$Z_{ChipMatched} = Z_0$

Impedance matching network

RFID Chip i

Z_i

Z_i

Figure 11.3 RFID chip control circuitry consisting of RFID chip, impedance matching circuit, and resonant switch.

intrinsic inductance (L_{res}) form an LC resonator tuned to the RFID operating frequency. We will refer to the circuit formed of the switch and capacitor as the "resonant switch." Figure 11.3 shows the RFID chip control circuitry consisting of the RFID chip, impedance matching circuit, and resonant switch. The equivalent input impedance of this circuit can take two values as in (11.1), and by connecting this circuit to an antenna one could build a simple one-button battery-less RC.

$$Z_{i=k} = 50\ \Omega, \text{for button } k \text{ actuated by the user,}$$

$$Z_{i\neq k} = 0\ \Omega, \text{for all the other buttons not actuated by the user.} \qquad (11.1)$$

11.2.3 Custom Switch and Multiplexing Mechanism

In addition to controlling the individual RFID chips, we require a mechanism to share the antenna of the RC device between multiple RFID chips. This was achieved by using a custom-designed SPnT switching network, which routes the RFID chip associated with the button being pressed by the user at each moment to the antenna while keeping all the other chips inactive. Figure 11.4(a) shows a diagram of the proposed switching network containing N access ports for the RFID chips and a shared port for the RC antenna. The termination impedance Z_i ($i = 1, 2, \ldots, N$) attached to each vertical transmission line resonator replicates the circuit in Figure 11.3. By observing (11.1) and selecting proper phase shift values for the vertical and horizontal transmission lines, φ_1 and φ_2, respectively, we can guarantee that:

- By default, all RFID chips are short-circuited and unreadable, and their transmission line resonators are terminated with a 0 Ω impedance. We refer to these chips as inactive chips.

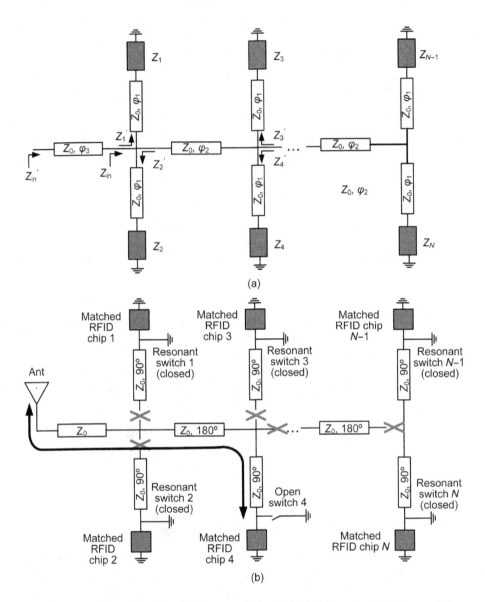

Figure 11.4 (a) Proposed SPNT switching network for N RFID chips and push-button switches, and a shared antenna. (b) Illustration of routing mechanism where the user presses switch 4 and the corresponding RFID chip gets routed to the RC antenna while all other chips remain inactive and "invisible" to the antenna.

- As the user presses button k, the associated transmission line resonator gets matched to the characteristic impedance of the SPnT switch ($Z_k = Z_0 = 50 \, \Omega$) and the corresponding RFID chip is routed to the RC antenna allowing the RFID reader to access its ID. An RFID chip in this state is referred to as an active chip.
- Inactive RFID chips are "invisible" to the RC antenna and do not interfere with the active chip.

Assuming lossless transmission lines, if the phase shift introduced by the horizontal transmission lines, φ_2, is equal to $n \times 180°$ ($n = 0, 1, 2, \ldots$), then the impedance seen into the shared SPnT antenna port equals the parallel association of all the impedances seen into the vertical transmission line resonators, $Z_{in} = [1/Z_1' + 1/Z_2' + \cdots + 1/Z_N']^{-1}$. Using $Z_i' = Z_0[Z_i + jZ_0 \tan(\beta_1 l_1)]/[Z_0 + jZ_i \tan(\beta_1 l_1)]$ leads to the equivalent impedance seen into the antenna port:

$$Z_{in} = \cfrac{1}{\left[\cfrac{Z_0 + jZ_1 \tan(\varphi_1)}{Z_0(Z_1 + jZ_0 \tan(\varphi_1))} + \cfrac{Z_0 + jZ_2 \tan(\varphi_1)}{Z_0(Z_2 + jZ_0 \tan(\varphi_1))} + \cdots + \cfrac{Z_0 + jZ_N \tan(\varphi_1)}{Z_0(Z_N + jZ_0 \tan(\varphi_1))}\right]}.$$

(11.2)

Finally, if we select the phase shift of the transmission line resonators φ_1 to be $90° + n \times 180°$ ($n = 0, 1, 2, \ldots$) and assume a single active RFID chip k with matched impedance ($Z_k = Z_0$) and inactive chips with $0\ \Omega$ impedances ($Z_{i \neq k} = 0\ \Omega$), then the equivalent impedance at the SPnT antenna port simplifies to $Z_{in} = Z_k = Z_0$.

Figure 11.4(b) illustrates the operation of the SPnT network. Assume that the user pressed button number 4, causing the switch to open and the corresponding transmission line resonator to be terminated by the 50 Ω impedance of the matched RFID chip 4. In this situation, chips 1, 2, 3, and 5 to N, which are shunted by closed resonant switches, remain inactive, presenting a $0\ \Omega$ impedance to their quarter-wavelength transmission line resonators, which then transform the $0\ \Omega$ termination impedances into open circuits. This way, only RFID chip 4 is routed to the antenna port while all the other chips remain inactive and do not interfere with the active chip as anticipated in (11.2).

The crosses in Figure 11.4(b) represent open circuits resulting from short circuits across the terminals of the RFID chips and lossless transmission lines. In practice, neither the short circuits nor the transmission lines are perfect. The SPnT circuit should present minimal insertion loss between the antenna and the active port, good isolation between the antenna port and the inactive ports, and reduced crosstalk between the active port and the inactive ports.

The horizontal transmission lines in the SPnT switch configuration of Figure 11.4 are only used for the purpose of increasing the freedom in the circuit layout design and have minimal impact for low-loss substrates. We designed our prototypes using microstrip technology for operation at UHF frequencies and fabricated them on low-cost FR4 substrate. The approach illustrated in Figures 11.3 and 11.4 is just one possible way of implementing the multi-RFID RC scheme. Alternative configurations are presented in Section 11.6 at the end of this chapter. For literature on conventional microwave SPnT switches, we direct the interested reader to [8–10].

11.3   Battery-Less RFID RC Design

This section details the design and characterization of the RFID circuitry, push switch, custom SPnT switch, and battery-less remote control prototype. The prototype is integrated into the demonstration system discussed in the next section.

11.3.1   RFID and Switch Circuitry Characterization

For our battery-less RC prototype, we used off-the-shelf ISO 18000-6-compliant RFID chips from NXP Semiconductors [11] and standard tactile push-button switches. We first characterized the RFID chips and impedance-matched them to the 50 Ω characteristic impedance of the microstrip transmission line system used in the SPnT circuit (see [12] for details on RFID impedance characterization and matching). We also characterized the RF response of the tactile push-button switches and found that even though they are not originally designed for RF, they presented satisfactory performance and repeatability for our application at UHF frequencies.

To perform scattering-parameter measurements on our chip/switch devices under test (DUTs), we used a vector network analyzer (VNA). For in-fixture calibrated measurements of our DUTs, we mounted them on SMA connectors and used custom calibration standards on similar connectors (see inset of Figure 11.5). Our custom one-port short-open-load (SOL) calibration kit comprised a short standard obtained by shunting the inner and outer conductors of an SMA connector, an open standard implemented simply with an open SMA connector, plus a load standard made of a high-precision 50 Ω resistor [13]. These custom calibration standards were modeled and specified in the VNA user interface software [14]. The calibrated scattering-

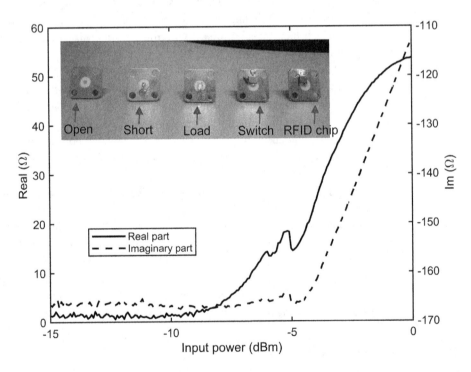

Figure 11.5 Calibrated measured impedance of unmatched RFID chip. The inflection in the real part of the impedance around −5 dBm corresponds to the minimum activation power level of the chip. The inset shows the DUTs and custom calibration standards mounted on SMA connectors.

parameter measurements of the DUTs were imported into the advanced design system (ADS) software for further analysis and design.

Figure 11.5 depicts the calibrated measured power-dependent input impedance of the RFID chip used in this work. Since commercial RFID chips typically present non-linear impedance, impedance matching should be performed at their minimum activation power, which is characterized by an inflection in the real part of the input impedance [15] (see Figure 11.5). Based on the minimum activation power level of the unmatched chip $(P_{unmatched} \approx -5 \text{ dBm})$ and reflection coefficient $\left(|P_{unmatched}|^2 = 0.86\right)$, we can use (11.3) to estimate the minimum activation power of the matched chip $(P_{matched} \approx -13 \text{ dBm})$:

$$P_{matched} = P_{unmatched}\left(1 - |\Gamma_{unmatched}|^2\right). \tag{11.3}$$

The measurements of the standalone push-button switch, resonant switch, and matched RFID chip in parallel with the resonant switch are presented in Figure 11.6. For the standalone switch, point A in the Smith chart corresponds to

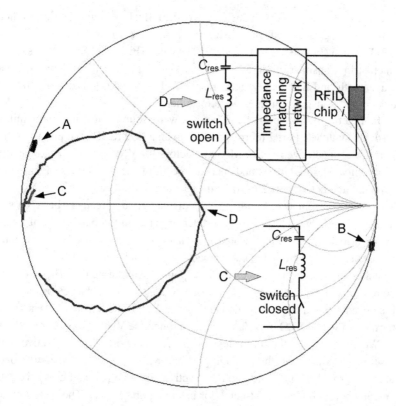

Figure 11.6 Typical measurements of the standalone push button switch (A and B), resonant closed switch (C), and matched RFID chip in parallel the with resonant open switch (D). These measurements were made in the 800 MHz to 900 MHz bandwidth. A simple L-matching network was used to match the RFID chip to 50 Ω [12].

the default closed position where the switch has a predominant inductive behavior, while point B corresponds to the open switch position where the switch behaves as a capacitor. In the closed position, the resonant switch presents low impedance at 866.6 MHz (point C), shunting the RFID chip and forcing it to the inactive state. Conversely, the resonant switch has high impedance in the open position and the equivalent circuit formed of the matched RFID chip plus the resonator switch is well matched to 50 Ω at 866.6 MHz (point D). In this state, the RFID chip can be accessed by the RFID reader. The series capacitance C_{res} is selected to form, in conjunction with the intrinsic series inductance of the switch L_{res}, an LC resonant circuit at the operating frequency f_c such that

$$C_{res} = \frac{1}{L_{res}(2\pi f_c)^2}.$$ (11.4)

11.3.2   Switch Design and Characterization

We used Keysight's ADS to design microstrip SPnT switches, fabricated them in FR4 substrate, and measured them using a VNA. Here we present results for a custom SP4T switch (Figure 11.7(a)) and SP10T switch (Figure 11.8(a)). The side and center microstrip transmission lines in this design have electrical lengths of a quarter wavelength and half wavelength, respectively, at 866.6 MHz. For a more compact design, the transmission lines were meandered.

In Figure 11.7(a), ports 1, 3, and 4 were shunted to ground to mimic closed resonant switches. Port 2 (corresponding to an active RFID chip port) and the antenna port are connectorized with SMA connectors for VNA S-parameter characterization. We compared simulations and measurements of the return loss of the antenna port (S_{11}) and the active port 2 (S_{22}), and the insertion loss between these two ports (S_{21}). The results presented in Figure 11.7(b) display good agreement between simulations and measurements, and present good performance in the bandwidth of interest.

We also designed and fabricated a 10-channel switch (Figure 11.8(a)) that had half of its ports shunted to ground to mimic inactive ports and the other half attached to resonant switches that allowed us to activate these ports at will during test. The switch ports were connectorized with SMA connectors for VNA S-parameter characterization. To achieve the impedance-matching condition (11.1), the shared antenna port and active port k were terminated with a 50 Ω impedance either through the VNA test port or an external load.

We characterized the return loss of the antenna port (S_{11}) and active port k (S_{kk}), insertion loss between the antenna port and active port (S_{1k}), crosstalk between the active port and the nearest-neighbor inactive port ($S_{k,k+1}$), and crosstalk between the antenna port and the nearest-neighbor inactive port ($S_{1,k+1}$). The measurement results are presented in Figure 11.8(b) and summarized in Table 11.1 for a carrier frequency of 884 MHz.

The SP10T switch shows acceptable return loss, insertion loss, and crosstalk performance across a 40 MHz bandwidth (see Figure 11.8 and Table 11.1). Even

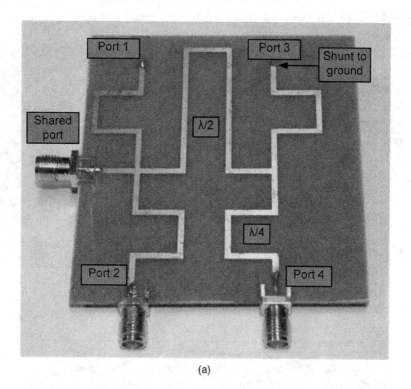

(a)

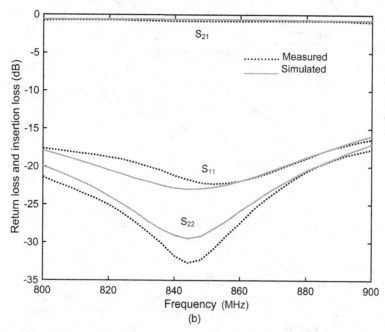

(b)

Figure 11.7 (a) SP4T switch prototype fabricated using microstrip transmission lines on FR4. (b) Simulation and measurement comparison.

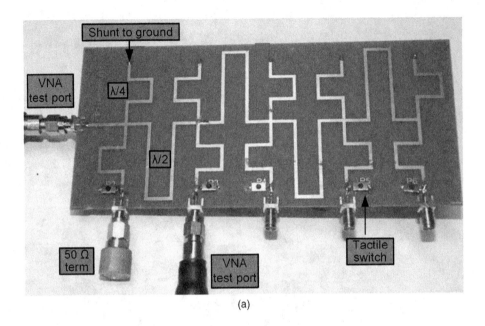

(a)

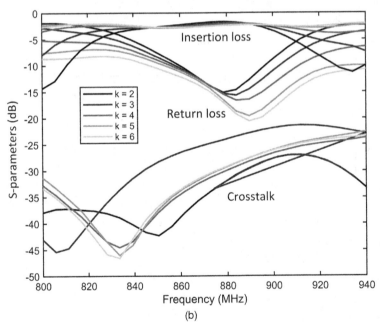

(b)

Figure 11.8 (a) SP10T microstrip switch prototype fabricated on FR4. (b) Measured return loss at the antenna port (middle), insertion loss between the antenna port and the active port k (top), and crosstalk between the antenna port and an adjacent inactive port $k + 1$ (bottom) as a function of the button k pressed by the user.

Table 11.1 SP10T switch performance at 884 MHz as a function of the active port k.

k	Return loss S_{11} (dB)	Return loss $S_{k,k}$ (dB)	Insertion loss $S_{1,k}$ (dB)	Crosstalk $S_{1,k+1}$ (dB)	Crosstalk $S_{k,k+1}$ (dB)
2	14.6	18.3	1.8	31.0	31.1
3	15.6	21.5	1.9	23.5	23.8
4	16.6	25.1	2.0	29.2	29.3
5	18.7	22.5	2.2	28.5	20.2
6	19.2	22.9	2.3	28.4	28.3

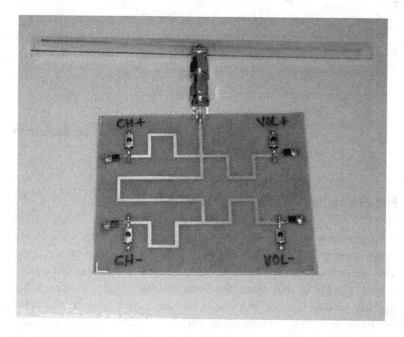

Figure 11.9 Four-button battery-less RC prototype with channel up and down and volume up and down functions.

though there is room for improving crosstalk (e.g., by using a better substrate and by optimizing the design), the performance obtained is sufficient for preventing unintended RFID chip activation due to leakage.

11.3.3 Battery-Less Remote-Control Prototype

Figure 11.9 shows a photograph of a four-button RC prototype that can perform four operations on a TV receiver, namely channel up and down and volume up and down. A simple dipole antenna designed for a center frequency of 866.6 MHz was attached to the shared port of the RC for interfacing with an RFID interrogator. Table 11.2 presents experimental communication ranges of the RC system as a function of RFID reader transmitted power. We achieved an operation range of 3.5 m for an

Table 11.2 Communication range as a function of the reader transmit power level.

Reader power (dBm)	Reading range (m)
17	1.05
18	1.22
19	1.39
20	1.90
21	2.24
22	2.43
23	2.53
24	2.61
25	2.98
26	3.22
27	3.50

interrogation power of 27 dBm and circularly polarized interrogator transmit and receive antennas with 5.5 dBi gain.

11.4   Full Prototype

To demonstrate the proposed battery-less control approach, we implemented a full demo where the RC prototype described in the previous section was integrated with a TV receiver (see Figure 11.10). For that, we used an off-the-shelf commercial RFID reader [16, 17] to interrogate the battery-less RC and a custom RFID-to-infrared interface module that acts as a bridge between the RFID reader and the TV receiver.

The custom RFID-to-infrared interface consisted of a universal infrared remote control whose buttons are operated via relay switches that are wired to a digital output port of the RFID reader. This approach allowed us to readily engineer a system prototype that could be adjusted to different TV receivers. An alternative to this approach could be to directly wire the reader digital output port to the terminals of the photodiode in the TV receiver and modulate the signal at this port appropriately. In a real application, the RFID reader system should be incorporated into the controlled device. To establish communication between the battery-less RC and the TV receiver, the RFID reader interrogates the battery-less RC to identify the button pressed by the user and then conveys this information to the TV via the RFID-to-infrared interface module in direct line of sight with the TV receiver (see Figure 11.10(a)).

The RFID reader was controlled via a Java application programing interface (API) provided with the reader [17]. A simplified flowchart of the developed application, which communicated with the RFID reader via TCP/IP, is shown in Figure 11.11(a). The RFID reader continually scans its interrogation field and when it detects a valid transponder ID, it sends a digital signal to the RFID-to-infrared interface to activate the

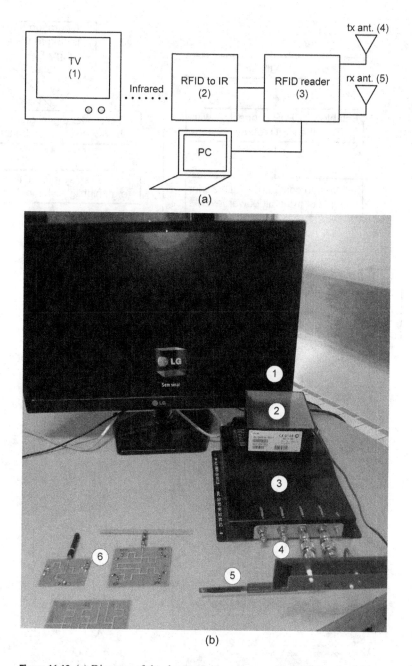

Figure 11.10 (a) Diagram of the demonstration system. (b) Photograph of the prototype (control PC not shown). 1: TV. 2: RFID-to-infrared interface. 3: RFID reader. 4: Transmitting antenna. 5: Receiving antenna. 6: Remote control prototypes. Note: For this proof of concept, we used a commercial TV receiver and off-the-shelf RFID reader but the battery-less remote control prototype and interface was fully developed in-house.

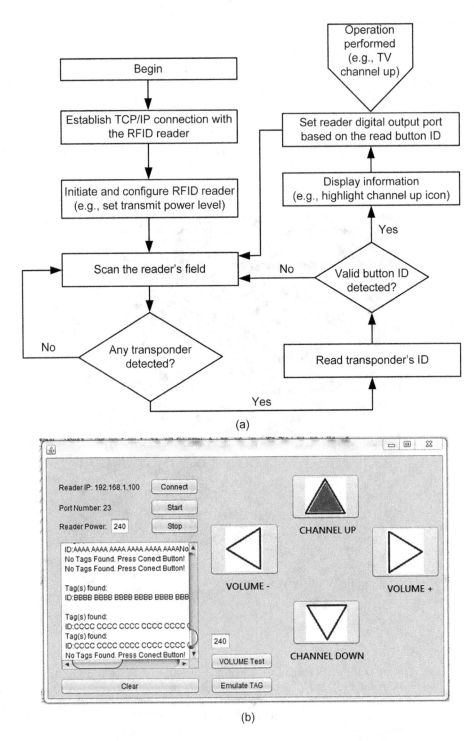

(a)

(b)

Figure 11.11 (a) Simplified flowchart of the RFID reader software application developed in Java. (b) User interface of the developed Java software application that controls the RFID reader.

corresponding button on the universal remote control. The developed application also provides a user interface (Figure 11.11(b)) for customizing various parameters of the reader and displaying the transponder ID, user control action, and other information.

11.5   Conclusions

We proposed a battery-less backscatter RC system where the RC device is wirelessly powered by the controlled device and backscatters the received RF signal to communicate data to the controlled device. In this chapter, we reported on the design and prototype of a battery-less RC system based on a multi-RFID scheme. Our prototype used off-the-shelf RFID chips and a custom SPnT microstrip switch with an activation mechanism to control the individual chips and share the RC antenna between them.

We built a four-button RC prototype capable of performing four basic operations, namely channel up and down and volume up and down, and we used an off-the-shelf RFID reader and a custom RFID-to-infrared interface module to integrate the battery-less control system into a standard TV receiver. Our prototype achieved a communication range of 3.5 m for a reader transmit power level of 27 dBm. Even though we demonstrated the concept on a TV receiver, it can be applied to other scenarios. Alternative and improved circuit configurations for the battery-less RC are presented in Section 11.6.

11.6   Alternative Battery-less Controller Configurations

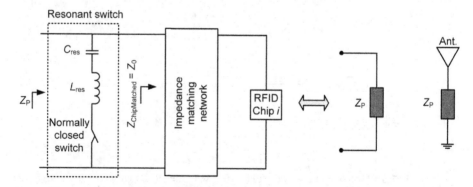

Figure 11.12 Parallel RFID chip activation mechanism based on normally closed switch. Left: RFID chip activation circuitry. Middle: Equivalent impedance. Right: Single-button controller.

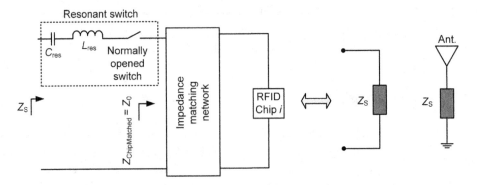

Figure 11.13 Series RFID chip activation mechanism based on normally open switch. Left: RFID chip activation circuitry. Middle: Equivalent impedance. Right: Single-button controller.

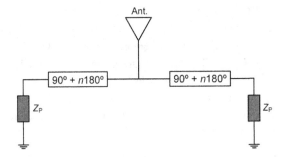

Figure 11.14 Two-button controller using the parallel RFID chip activation mechanism of Figure 11.13.

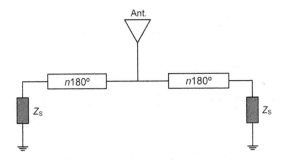

Figure 11.15 Two-button controller using the series RFID chip activation mechanism of Figure 11.14.

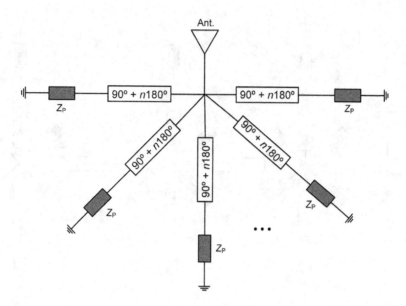

Figure 11.16 *N*-button controller in star configuration using the parallel activation mechanism.

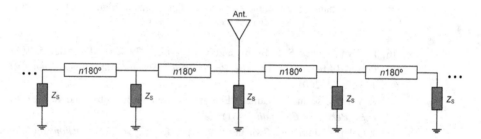

Figure 11.17 *N*-button controller in cascade configuration using the series activation mechanism.

References

[1] Engineering and Technology History Wiki, "Biography of Eugene J. Polley." [Online], available at: https://ethw.org/Eugene_J._Polley

[2] E. Olivetti, J. Gregory, and R. Kirchain, "Life Cycle Impacts of Alkaline Batteries With a Focus On End-of-Life." Massachusetts Institute of Technology, 2011.

[3] European Portable Battery Association, "The Collection of Waste Portable Batteries in Europe in View of the Achievability of the Collection Targets Set by Batteries Directive 2006/66/EC," 2016.

[4] S. Dong-Feng et al., "Research of New Wireless Sensor Network Protocol: ZigBee RF4CE," in *International Conference on Electrical and Control Engineering*, 2010, pp. 2921–2924.

[5] A. Nechibvute, A. Chawanda, and P. Luhanga, "Piezoelectric Energy Harvesting Devices: An Alternative Energy Source for Wireless Sensors," *Smart Materials Research*, 2012:853481, 2012.

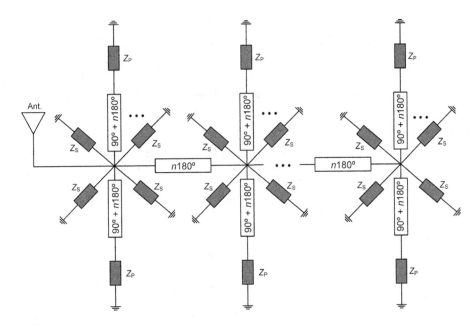

Figure 11.18 *N*-button controller in a hybrid configuration using both parallel and series activation mechanisms. Here, the previous configurations of Figure 11.14, Figure 11.17, and Figure 11.18 are combined for increased button density.

[6] P. Glynne-Jones, S. P. Beeby, and N. M. White, "Toward a Piezoelectric Vibration-Powered Microgenerator," *IEE Proceedings – Science, Measurement and Technology*, 148:68–72, 2001.

[7] J. A. Paradiso and T. Starner, "Energy Scavenging for Mobile and Wireless Electronics," *IEEE Pervasive Computing*, 4(1):18–27, 2005.

[8] J. Galejs, "Multidiode Switches (Correspondence)," *IRE Transactions on Microwave Theory and Techniques*, 8(5):566–569, 1960.

[9] J. F. White and K. E. Mortenson, "Diode SPDT Switching at High Power with Octave Microwave Bandwidth," *IEEE Transactions on Microwave Theory and Techniques*, 16 (1):30–36, 1968.

[10] D. J. Kim et al., "Switched Microstrip Array Antenna for RFID Systems," *Proceedings of the 38th European Microwave Conference*, 2008, pp. 1254–1257.

[11] NXP Semiconductors, "SL3S1002 RFID chip, UCODE G2XM and G2XL, Product Data Sheet." [Online], available at: www.nxp.com/docs/en/data-sheet/SL3ICS1002_1202.pdf

[12] P. V. Nikitin, K. V. S. Rao, R. Martinez, and S. F. Lam, "Sensitivity and Impedance Measurements of UHF RFID Chips," *IEEE Transactions on Microwave Theory and Techniques*, 57(5):1297–1302, 2009.

[13] Bourns Resistive Products, "CHF1206CNT resistor." [Online], available at: www.bourns .com/docs/products-general/Bourns_Resistive_Products_Overview_White_Paper.pdf

[14] M. Hiebel, *Fundamentals of Vector Network Analysis*, 1st edition. Rohde & Schwarz, 2007.

[15] R. Kronberger, A. Geissler, and B. Friedmann, "New Methods to Determine the Impedance of UHF RFID Chips," in *IEEE International Conference on RFID*, 2010.

[16] Alien Technology, "Hardware Setup Guide ALR-9800," 2008.

[17] Alien Technology, "Reader Interface Guide, All Fixed Readers," 2016.

Index

Printed in the United States
by Baker & Taylor Publisher Services